THE GALACTIC FEDERATION

DISCOVERING THE UNKNOWN CAN BE STRANGER THAN FICTION

STEPHANIE P. WALKER

Tellwell Talent
www.tellwell.ca

ISBN
978-0-2288-7826-1 (Paperback)
978-0-2288-7827-8 (eBook)

Top &Side View/Sketch

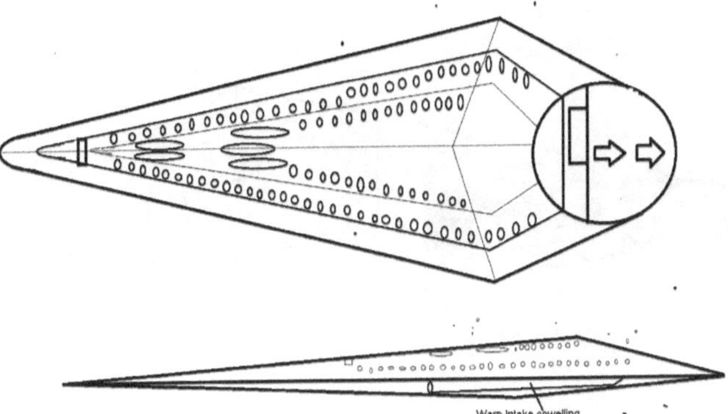

Deck #1 Engineering

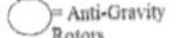

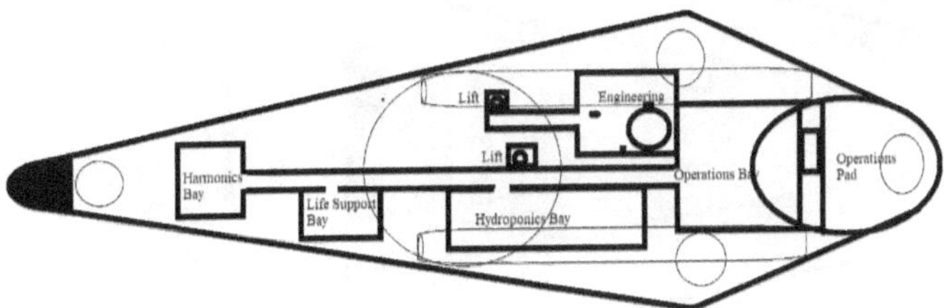

Deck #2 Habitat Deck

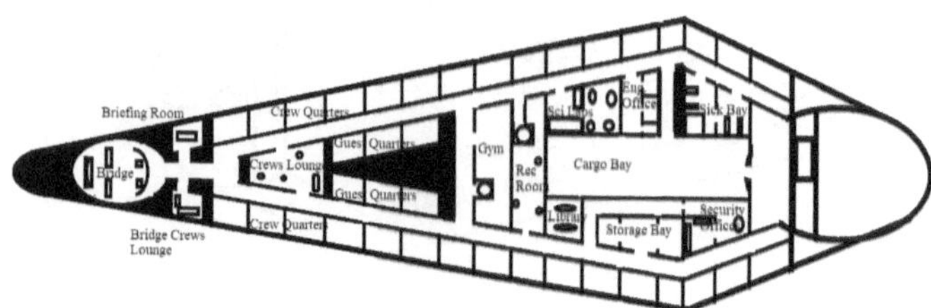

Deck #3
Observation Deck

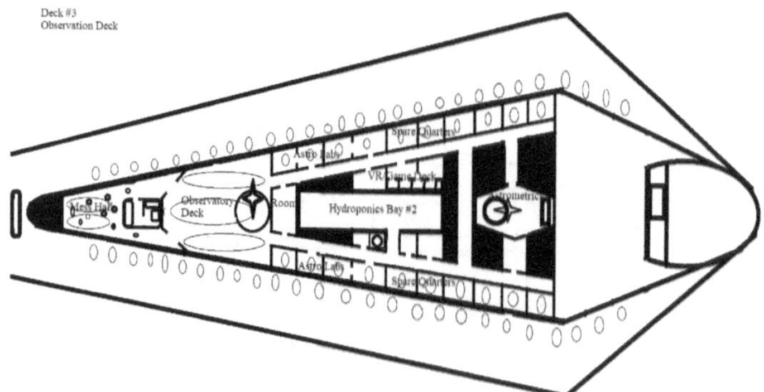

ACKNOWLEDGEMENTS

First of all, I'd like to thank my imagination because without it this series would not exist. Second, I'd love to thank my mother for giving me and nurturing my overactive imagination and putting up with everything. Mom, this book is dedicated to you! You get to see me shine, so here it is. My mom supported this project and believed I could it, and now here it is. I thank my in-laws for their support as well as my few golden friends in life. Thank you for listening to me ramble on and on about this book for years, and for your kindness and support when I had questions regarding reader satisfaction. Thank you to Gene Roddenberry and the *Star Trek* captains and crews for your acting and storytelling. You have inspired me to create this series. RIP to our fallen Trekkie actors. Your good times on TV will never be forgotten and are missed.

Last but not least, a huge thank you to all my soon-to-be fans. I hope you all enjoy the journey ahead, and maybe one day we'll be lucky enough to work onboard a starship exploring our galaxy together.

PROLOGUE

In the summer of 2019 in the Fraser Valley, the small town of Chilliwack witnessed an event that defied all logical explanations. Not only that, but two students from the local university astronomy department discovered unusual and curious activity flying around the Pleiadean system some four hundred light years from Earth. The two-month study, tapes, and constant surveillance all showed signs of active and intelligent light movement coming from that system. It was heralded as quite the discovery on local television for a few days before it was killed off and officially reported by space agencies as just light refractions. But clearly this wasn't the case.

The two astronomy students were told by the local military authorities not to report such matters anymore, yet locals and people in other parts of Canada thought differently, including the university's Sciences and Aerospace Departments. A Hopi male chief from an Arizona tribe named Ahote spoke to the university and the students and locals about a "spiritually advanced people who came in the old days and saved my people from earth changes. They are humans just like us and look like us, and they reflect the many races we see on Earth. They are the Pleiadeans from the Pleiades system in the constellation of Taurus."

It first started out as a joke, but the university's Engineering Department claimed they would build a spacecraft if Ahote could prove the Pleiadeans existed. He told them to build him a spaceship so they could go there and he could prove it. Then the idea became less funny and more serious. Slowly students from the university's Sciences, Engineering, and Aerospace Departments were all missing after their day's studies, doing, as one claimed, "extra credit for a group class. It's a really, really big project."

Five months later, as described by local residents who witnessed the event, "There was a flying oval shaped thing with a red light trailing behind from two rocket-looking things … except it wasn't rocket exhaust, and the craft was silent." The local Chilliwack Police Department also saw this craft, as well others who were in pursuit of it briefly as it took off from the farm and headed north before the locals reported the sightings. "But why?" the town asked. The police wouldn't say anything, but it was known that they were storming through town in a convoy with silent lights minutes before the raid on the Braxton farm began.

The parents of the students said that their kids were immersed in some after-school extra credit project that was huge, and it was taking place on the Braxton family farm. For a year they synced their classes together and then went to school in the morning. They got off at three in the afternoon and didn't get home until three in the morning. They'd sleep for a few hours and then get up and do the same thing over and over again. But on the day of the event, the parents and some university staff were told that an open invitation event was being held on the Braxton family farm in Rosedale, and only an invited few were invited for some unveiling. They explained that something spectacular had happened, and it was only a matter of time before they could share it with whoever would be able to see it, and it was going to be epic. The parents thought their youths were up to something big and had a part in what happened that day.

CHAPTER ONE

Braxton Family Farm, Rosedale, Chilliwack, British Columbia
Earthdate: 08192019, 1:30 p.m.

Kate Braxton sat in her new leather chair for the first time ever.
It was her captain's chair, and it was new. Everything was new in
the ship they'd just built, including the circular bridge she was
sitting in with its light grey carpeted walls, windshield, and the
pilot's console that sat in front of it and seated the ship's pilot and
communications officer. To her left she looked at a similar chair,
which was the commander's seat, and then a few more engineers'
consoles along the far bulkhead with wallcomps. To the right sat
the ship's science officer and the ship's defence officer's console
and wallcomps. Every console and all wall computers were touch
tone key and dial display that you interacted with like the laptops
of today. The engineers' consoles could do what the engine room
computers did, but ship's chief engineer, Elisa Torres, liked the
idea of being on the bridge and having the same access terminal
with the captain and commander, even if the engine room was out
of service due to an emergency. The back bulkhead behind them
had a built-in computer screen that served as face-to-face audio-
visual contact with others on a FaceTime-like app, but a corridor
split through the back bulkhead that led to the lift, the briefing
room access and bridge, and the crews' lounge and coffee bar. A

railing encircled half the bridge and an Epad console between the captain's and commander's seat. The ship's lighting was perfect! The adjustable mood lighting provided a soft, bright light that softened the ship's atmosphere and brightness day and night.

Kate lay back in her chair, legs crossed, wearing her navy blue flight uniform, which was a military consignment shop Air Force ground flight suit with a homemade *Valkyrie* flight patch on her left shoulder that a friend made for the crew uniforms. As she read the progress reports updated by all crews who helped build this ship, the C.S.S. *Valkyrie*, she noticed that the ship's chef had reported all food and water was on board and stowed and the galley and mess area. They were ready for departure that night. Kate smiled at this, as the chef was the university's coffee shop owner, whom some students saw as a second mother or grandmother. She suddenly looked up to hear soft but heavy footsteps approaching from behind. She stood up and turned her head to see Ahote, the first officer, exit from the hydraulic lift. He walked up to the railing and peered down at Katie with a smile.

"Captain, do you have a minute?"

"Commander, I was just going to find you; come join me." She motioned for him to join her on the bridge.

He smiled warmly and walked around the rail and sat in his seat next to the captain. He leaned his head toward her, making eye contact, before she went on. "I've read the updates and I'm happy to report that the mess and food stores are all ready. The security teams are together and report that they're ready around the farm. Elisa Torres is in Engineering at this moment with Pierre, running tests on our new warp drive system, so it looks like we're hours ahead of schedule."

Katie Braxton was the *Valkyrie*'s captain at 5' 7" with natural straight, shoulder-length, layered, auburn colour hair tied back into braided ponytail. She was Caucasian with green eyes, a small pointy nose, and an oval British-looking face with a slim and partly athletic body. She was naturally pretty good looking but had a fiery temper that ran as fierce as her face looked when angered or challenged in a threatening way. Ahote, her first officer, was 6' 2" with long, slicked back, black hair, brown skin, a soft looking roundish head, and a manly looking native complexion, as he was a Hopi Indian from a potato farm in New Mexico. He had friendly hazel eyes and a strong, medium build that made him tall and handsome; he was thirty-six years old and wore the same flight suit as Katie, with the same colour and patch on the left shoulder. This was the uniform for the whole crew.

"Well, this is all very good news, Commander. This mission just may be a go after all." She looked at the commander as she said this, her eyes full of excitement. They both chuckled with disbelief that they'd just pulled off something secret and impossible and extremely dangerous, as this was a big part of the plan. The commander nodded his head in agreement of their flight plan that took them over the town of Chilliwack for all to see, and then it hit him that when Kate said many months ago that she was going to make this flight for all to see, she actually meant it, because now he was looking at the orders on his Epad.

It was a joke at first, but one year ago, after the university's new Sky Observatory and Science Club had spotted some strange, intelligent, controlled lights flying around the Pleiadean system, Kate and her friends in the university Aerospace Engineering Club came to the conclusion that if they could get materials, the crew could build a spaceship. At first the discovery was taped and documented, and hours and hours were put into looking at the footage, which was found to be genuine and not a fake. This discovery went viral for a while before being shut down and dismissed. Soon after their engineering professor, Pierre Gagnon, overheard a little of this plan and claimed he knew how to make a warp drive engine and that he had the schematics he'd hacked from the National Aerospace Association department of the last university he'd worked at. That was another story on its own— ohhhh geez! Somehow Gagnon got away with it. He had said he would build a warp engine if the students were to build this ship and get the necessary materials.

Then came a random science student who expressed interest in joining up. He introduced himself as Kyle White. He had a cousin who was an aerospace engineer for the Canadian government, a "secret ally" who was willing to present them with a transparent metal not yet disclosed to the public. They used it for the ship's windshield and windows. Kyle also explained that he could build an anti-gravity generator and that there was a way to play with the planets' natural magnetic energy fields to use energy in space to propel this ship. He could for sure get the materials. They all thought he was crazy, but in a few weeks, everyone involved was eating skepticism when they received a call from his friend, who had the materials ready to ship out to the students.

Next they needed a reason to build the ship for real space travel, and then they needed a destination. Ahote had told of an ancient people from the Pleiadean system that visited his people thousands and thousands of years ago; he had said that his people had built a small village in the same configuration of the Pleiades system in New Mexico in the deep desert. It was funny that he brought that up, as for three years now people had been reporting UFO sightings with contact experiences with advanced humans like us who claimed to come from Pleiades. They said that these aliens always warned the contacted about our governments and humans eventually arriving in space; of course, no one could put two and two together, so everyone thought it was crazy. But as the years passed there were more sightings, and people started to really wonder.

But now there was a reason to build the Civilian Spaceship *Valkyrie*! These students were going to travel four hundred light years to the system to see if the Pleiadean's home worlds exist, and then they'd come back to report it to Earth's people. But first they had to conduct a test flight and reveal this vessel to their families and town without government interference and agencies freaking out. Ahote sure believed in them, and Kate trusted him and his visions and native spirituality, as in the past he was never wrong about anything. Ahote was the tall, silent-but-wise looking man who you knew was intelligent and he looked it and felt it too, as it radiated off him like an energy field of some sort.

For the last year, twenty or so of these students and their friends had come together to secretly build this ship, with "unknown" help, in the woods behind Kate's family farm. There was a cleared out section, one thousand by eight hundred feet, with a huge green tarp and mesh netting overhead to hide the ship from above a few yards into the dense forest behind the farm. Here they built in secret with no problems or interruptions. But then came the Secret Service raid at Pierre's house, and they took his blueprints for the warp drive core. Then he lost his job at the university. He was almost tried in court but the case was thrown out due to his voluntary silence on the matter and his promise to stay away from such materials. From there he was tossed to the curb and the engine build was stalled until further notice. A few days later, he showed up at the farm saying that the plan was still on and he had a second copy of the schematics. From then on he lived on the farm while working full-time undercover, while the Chilliwack University board thought he'd disappeared for good.

Finally, today was the day for the last touch-up on *Valkyrie*. Now she was complete, with all systems functioning at 100 per cent in standby mode. Engineering crews all reported that their systems' power fluctuations and algorithms were at a steady level and holding, and science crews reported that the ship's electromagnetic hull was ready and currently absorbing the planet's energy currents that were being syphoned and run to the harmonics centre. This would help propel the ship off the ground with its anti-gravity rotor. The syphon generators were now active and charging the reserve power units. Today was the first and only test flight they had planned, and all of Chilliwack would witness it. After it ended, hell knew what would go down upon arrival.

The *Valkyrie* was one thousand feet in length and eight hundred feet broad with an oval shaped saucer hull and huge, fat tubed Nacelles intakes on each side under the hull, built into the hull for sleekness. Ahote and Katie sat there laughing and talking about the successful construction of the ship and how a year before they had scoffed at the idea and laughed in disbelief, but now they were sitting on the bridge of the ship they'd built.

The commander's eyes widened as he took in a breath and spoke up. "So I've been studying the flight plan Ms. Blackwood plotted and sent to me."

"Hmmm, she plotted a course?" Kate asked, looking at him with interest.

"Yes, with one of the astrometrics officers I believe," he said in his deep, wise tone. He explained that Rachel Blackwood and crewman Perez had gotten together to plan a trip from Earth to the moon and back at warp1, which would be a day trip at the very least before they returned to earth and landed in Kate's back yard. Kate giggled and commented on how cleverly she wanted the town to see this, so he was to plan a course that would stir up attention.

"Shall we go to the mess hall and join the rest of our crew in celebration, Captain?" asked Ahote, chuckling as Katie studied the flight path on an Epad that the crew used to write up reports or texts, or just communicate. It was like a tablet cell phone. She looked up and smiled then told him, "Lead the way, Commander."

Elisa Torres, a Chilliwack-born girl of Spanish decent (her mother and father were born in Spain) was 5' 5" with light brown-fair skin, a roundish head, and soft but troublesome looking groomed features. She had mid-neck-length black curly hair, blue eyes, and a slim build with an athletic tone to her body. She stood at her console next to the warp core preparing to do a ship-wide self-diagnostic while the other crews were heading up to the mess hall for celebrations. Pierre walked in, crept up to her, and handed the chief engineer a silver mug of coffee.

Startled, Ms. Torres jumped back a bit and looked at the mug in shock and then awe as she grabbed it and quickly took a drink of it. "Ohhh god, thank you so much. I'm too busy to go to the mess right now." She said this while looking up at him. Elisa Torres was pretty easy to get along with until she got mad or an attitude; her face would show it and she was a hot-headed bad ass when she got angry. But when she smiled, it was the funniest smile ever. She always looked happy and pretty but very mischievous too, and she didn't really like that about herself. She was so happy to learn that she was selected to do this mission. As secretive as it was, she had confidence and knew the material enough to know she was capable of making her part happen. Now here she was, leaning against her console, sipping a cup of coffee, and talking to Pierre as she explained the results of her system scans.

He chuckled as he stood there in his blue uniform; Pierre was 5' 9" with a broad build. At thirty-five years old, he had a tough looking French man's face but was really a compassionate educator and crazy scientist. He was the one with all the alternative energy ideas and technology, while everyone just brushed it off. He had short, brown, wavy hair and brown eyes. His friendly smile made him cute—in Ms. Torress' mind at least. His voice was kind of raspy but deep and very French sounding. When he talked to you, you knew you were communicating with an intelligent person who was too ahead of his time.

Mr. Gagnon smiled and said with commanding discretion, "Why don't we go to the mess hall and join our comrades in celebration for real now?" With that, he tapped her shoulder with his EPad, turned around, and walked into the corridor, tucking away his Pad. Elisa stared at him for a few seconds before coming to and jogging to catch up with him; together they walked straight until they got to the lift and pulled the latch to go up to Deck One.

"So how are you feeling about this test flight? Nervous?" she asked.

"Ohhhh yes, very nervous, but I can see the algorithms are in sync with all systems, so I'm quite positive also," Pierre replied as the lift reached Deck One and stopped with a clicking sound.

The two engineers stepped out of the lift and into a larger corridor. They turned left, walking and talking about how they were excited for the trip and quite sure everything would be ok. Turning the corner to the right, they met up with Kyle White, who was also on his way to the celebration. He looked up and flashed them a smile and a nod. "Hey hey," he said, "test flight coming up. Right on! Think we'll live?"

The engineers smiled back as Elisa spoke up. "Hey hey, yea, and it's our asses on the line should anything go wrong!" She shook her head and grinned as the two men's faces went blank for a second. Then she snorted in laughter. The guys broke out chuckling as she continued. "No need to feel bad, we're here because we're the best at what we do, chosen by others who trust us and our knowledge, as they work next to us. Hey hey, we better be sure everything works or we're space dust, but great job anyway, guys."

Moments later, they walked into the mess hall to see the room crowded with blue-uniformed crew members who were also local university students who ventured into this project. There was classical music playing somewhere, and the crew held champagne glasses, cheering and applauding each other. They were laughing and shouting congrats, and others were dancing or off in groups going over their Epads. Some just stood off to the side in pairs, watching the festivities, with coffee in their cups. The three were caught by surprise when they looked up and saw the three main observation windows that stared out forward. They were at least thirty feet in length. The room was fresh and new, with light grey walls and a café-style galley and counter. A little home in space goes a long way, said Cassidy, the ship's cook and hydroponic food grower. She wanted a little comfort too, as she'd be staring out those huge windows most of the time, watching the stars fly by. She had the greatest view, just like the bridge had. There were small, round tables everywhere just randomly littered throughout the room, with small vases with red roses in them set on top. There were also nice, comfy cushioned chairs. The tables were ordinary black university café tables, and Torres counted at least thirty of them. The room itself was pretty big, with a slanted ceiling that followed the ship's forward almond hull slope. The crews were having fun, and Cassidy was also having a blast as she poured drinks for the captain and the commander. They clinked glasses with her, sipped the drinks, and moved on into the crowd.

Kyle White stood at the fore windows looking out into the woods in front of them. He could see the security teams outside and a few cracks of sunlight breaking through the hidden tree and netting canopies; indeed, they were well hidden, and he admired this idea from Kate. Just hide and build the *Valkyrie* in plain sight pretty much, in a densely forested part of the farm property. It worked for one year and was still hidden, and now *Valkyrie* was completed. Kyle knew what Kate was thinking when she first approached him with this idea, and he was the first one to fully know this whole plan was possible. Not only that, but deep inside he knew what they were up against and what was up there waiting for them to find. He gazed out the window, wondering if anyone there knew or had any idea what they were getting into. But it was up to them to follow their hearts, even if it led to this moment. But that's not to say that he thought himself better than everyone else, because he did not. Yet he too was picked because of his science background and his new job offer from NASA. He indeed thought that Kate and this crew were brave souls whose thoughts and wonderings drove them to dream and then to create this ship, which was now ready for a test flight never before accomplished by civilian minds. They'd made it this far, they'd built this, and this crew had done all this. He was impressed.

Kyle had chosen to side with Kate and this group of people over his NASA job offer because he knew they would go places not yet imagined, and they had a plan and an idea of how temporal drive and warp technology worked. Kate showed the typical human character trait of being adventurous and curious. She believed she could build this ship if she had the materials, which he had, and the students' undying demand to build it and do it. He wanted to share that with them because they deserved it and he was there at the right time. He was truly destined to help them out. He also believed in the triumph of the human will, and this was their time to do the impossible and meet their destiny in the stars as Earth humans. But if they ever knew his deepest aspirations, they'd probably question his choice to hold back. Why should he spoil the surprise when they'd worked so hard to get this far; they should enjoy and learn, and just maybe he might intervene when the time came for them to rightfully face the reality of this universe.

He smiled as he thought about the success he foresaw for the trip, but he also knew that the ship couldn't be kept secret for too long. Sooner or later they'd have to hand *Valkyrie* over to the feds—or not! It was going to be a toss-up soon, and there was some chance most of the crew would chose to keep *Valkyrie* and be free to come and go as they pleased. A small minority might choose to jump.

"Damnit! Shut up, White. You talk too much; it's unsettling," he scolded himself. This mission could and would lead to bigger and better opportunities for all in the end; just how its played out would be the verdict. Who knew where the latter would take them?

Kyle White was 6' and slim but fairly well toned to fit his figure. He had a natura baby face with short, wavy, brown hair, blue eyes, and a European complexion. He had a typical build for a high school football captain, but his ego had been dulled a very long time ago, and his intellect and compassion and love of exploration in life led him to where he was now. He had an unearthly look and emanated a bright feeling full of spirit and vibrancy, yet he hated how women looked at him, not seeing him for what made him, him! His intellect and curious yet friendly nature was clouded over by a look of "elsewhereness." One could say he looked very stiff and reserved, with a connection to somewhere else and not here in the moment. But he knew that this was his time to play his part and be the crew member he had assured his peers he would be, no matter what they had asked. There was that need of oneness he felt lacking in most women, and for once he wanted to be known for his strengths and brains and not just his brawn for one drunken night of whatever.

Kate approached from out of nowhere with a smile on her face, holding a glass of champagne. Kyle smiled back, held up his glass, and drank with the captain before turning to her and shaking her hand in congrats of this accomplishment. "Captain Kate Braxton, it's a pleasure to be here, I must say again, and I thank you for this opportunity. This is indeed very exciting, and I'm honoured to be serving under you." He smiled as Kate replied.

"I reached out to the very best." With that, she raised her glass again.

"I know that this trip will be an adventure with positive results, Captain. I know of you and your life experience, and I know from personal experience and that of others that you're well qualified and knowledgeable of captaincy protocols. You grew up spending every summer on your grandfather's boat, helping out and running the crew and captaining. You demanded strict detail on the blueprints of this wondrous vessel, but you were also driven to learn as much as you could about how this all works in the short time you had. I'm all yours and reporting for duty." He smiled and nodded.

Kate smiled back and remarked thoughtfully, "Yeah, do you ever recall the time I was sent out with another boat captain as a first mate, and he turned out to be abusive to our crew, so I threw a mutiny and had him tied to the flag mast for a day before returning home to kick him off the boat? How the heck do you know about that part of my childhood?"

"Ahh, like every investment, Captain, one must be investigate or research before dropping funds to people or groups. I love what I concluded about this crew's ambitions," he quickly responded boldly with a confident and sophisticated tone. "Well, I guess this ship is in great hands then, and this is guaranteed to turn out in our favour. But just like on the sea, respect the environment around you, as space can take you out when it sees fit, I'm sure."

They spent the next thirty minutes troubleshooting possible scenarios to be expected of space travel, including the harmonics frequency changes he would be making as they hit different parts of space and approached the moon. Other planetary systems would also need to be adjusted in the ship's systems due to gravitational frequencies. Space travel really was a job not be taken lightly. "Have you got your flight path plotted and ready?" He brought this up to ensure that the ship had a hydroponics bay. He commented that it was certain that if other humans were in space on large ships, they too would have hydroponics bays on board. He really was amazed at the planning and research that had been put into the project.

Kate glared out the window as a few streams of sunlight shone through the tree canopy and into her eyes. She stepped back a bit and asked how to make the windshields automatically darken when bright light shines through, and Mr. White explained that he would look into that as the ship's first project if the ship wasn't whisked away by the feds upon its return. They both laughed and stared out of the fore window as Braxton sighed and whispered out loud, "Gosh, it would be a shame if we were to lose all this."

Ahote was walking around admiring all the crew having a good time talking to their colleagues and sipping their drinks. He was looking for someone from his class back at the university. She was a First Nations woman from the town who'd started talking to him last month about native lore, and he knew she was around somewhere. He'd forgotten her name, which was a little embarrassing, he thought. Walking through the crowds, he scanned faces but didn't see her. *Weird*, he thought, as he was told all crew were accounted for, but then he thought she might be in the crew quarters or a work station. Who knew, but he was determined to find and talk to her.

Valkyrie's Bridge

Rachel Blackwood, ship's pilot, and Thomas Hilt, communications officer, sat in their seats prematurely, obviously excited to get to work when the test flight came up. Both sat at the console while Rachel taught Tom a little more advanced piloting skills. She had studied the ship's controls and the planned navigation systems, flight systems, and helm control to make control of *Valkyrie* like flying a Boeing 737. But it was even more complicated when it came to turning the ship, so the best they had was negatively charged ion thrusters that reacted with the matter outside in space and pushed the ship in the opposite direction. At lower speeds, the vessel turned sharper. She finally got to the point were she could memorize each procedure in her head and then teach the auxiliary pilot, Thomas, for when he would be needed. He was just the communications officer by trade, but why not know a little bit more about piloting, especially when you'll be sitting next to one?

Rachel was transgender, male to female, and in her late thirties. At 5' 7" she had an athletic build and strong, oval shaped head with natural features. She had a small, crooked smile and soft but determined green eyes and natural red hair, layered and shoulder length. She was very smart and always eager to help out anyone who needed assistance, yet there was something about her that most people second guessed yet never put two and two together. She was the kind of girl who had natural beauty and big brains, and she was always there when someone was down on their luck. Rachel was a rebel to the core, as before this she'd been in the Air Force and flew military airliners full of troops abroad, but she was kicked out when she introduced a new manoeuvre that her flight instructors called reckless and careless, and in a 23 million dollar aircraft funded by the tax payer. She was hauled into the air marshal's office and stripped of her wings. Not only that, but she had questioned the military's stance on a few other policies, which led to a gag order a few times, but that never stopped her. Among these things was the whole bar fight incident with some local rednecks, but being one of the best pilots there, they had to keep her as long as they could until she was eventually booted. She went to the university for aviation structural design and upgraded her pilot's licences.

Thomas, on the other hand, was just a lax country boy of thirty standing 5' 9" with short, wavy, red hair. He was Caucasian with a slim, muscular build and a soft face. He was very handsome with blue eyes that looked brighter in some light. Thomas was a pretty relaxed and laid-back guy with an attitude to the point of laziness sometimes. His last job was as a police dispatcher until he went to the university for a technical degree in communications and was spotted by Kate and invited to come to the farm. There was an opportunity there for him—if he wanted it, of course. Not wanting to throw any opportunities away while keeping an eye out for something better than his current endeavour, Thomas was intrigued by Kate's offer and went along with it. When he first saw *Valkyrie* he laughed and was in total disbelief, but then he was given a full tour of the ship's systems. Well, let's say he was sold on this plan.

Now he was sitting on the bridge in his seat, learning some tricks to warp drive piloting. The two laughed as she explained how to turn the ship to her port or starboard with the thrusters that took *Valkyrie* into a small to large radius turn in a small amount of time. It was something Rachel thought was neat and explained that she was already thinking of a manoeuvre that could take them ninety degrees in seconds, and while she was at it, there should be backup manoeuvres too, thus putting the ship into controlled spins and what she called warp outs. They both laughed as the pilot excitedly spewed her ideas.

"So, I'm thinking that with a planet's magnetic field, harmonics could rig a syphon system to the pilot console where I could set up negative charges on the opposite side buffers and use it to rotate the ship like a spinning almond," Thomas laughed the replied.

"Yeah! Hey, that would be pretty sweet, actually."

"Then I think I'd like to make *Valkyrie* do donuts while engineering vents warp plasma," Ms. Blackwood said, giggling as Tom kept laughing. Thomas looked at Rachel like he did quite a bit since he'd been brought onto the ship for systems training. There was something about her he could quite pin, yet he thought her a perfect looking creature, not hung up on her looks but trying at least to look better with little makeup. She kept herself real, and everything she did was efficient. He couldn't explain it, but she had a regimen about her that he actually found attractive yet mysterious. He liked mystery because it was a challenge, but he also liked the surprise at the end. It obviously wasn't for everyone, but it was for him.

He found piloting *Valkyrie* very appealing and agreed to be an auxiliary pilot when Ahote suggested that they would need another pilot, perhaps a part-time one, just in case. Thomas jumped at this chance, and Rachel agreed to teach him the basics of warp travel and the console systems, especially how *Valkyrie* turned, which seemed to bend a lot of people's minds, as planes have rudders, and ailerons and ships and submarines also have rudders. As far as the *Valkyrie* engineers knew, it appeared that thrusters were the way to go, and piloting *Valkyrie* was just an art in itself waiting to be unlocked.

Out of the blue, Thomas asked Rachel, "Hey, do you think we'll run in to these Pleiadeans? Or any other life out there?"

"Sure, why not?" she replied. "Our universe is expanding; scientists have already discovered we live in a multiverse of times, space, and dimensions. Why not Pleiadeans?" she responded.

Crewman Tom was convinced that there was probably other life out there and that this reality we live in was way more exciting than most people knew. Just doing what they were about to do was considered science fiction, but look now! He had a feeling that if humans could get into space and travel this way, especially novice civilians, then there definitely had to be intelligent life elsewhere, as this technology couldn't be just from earth. Then he asked another question. "You're an aerospace engineer too, right? You helped Captain Braxton design this ship, right?"

Rachel smiled and then looked at him. "I did indeed, but most of the ideas were hers. I just put little things here and there together on paper." He nodded, intrigued at what she knew about this ship and about designing ships. She continued to explain how the captain was thinking of building another ship if this Pledan civilization did exist, thus opening up a chance to start a civilian space program between civilians of Earth and the Pledans. This really fascinated Tom, as he pictured himself the main pilot of such a ship. Then he thought he'd rather be a pilot in some space fleet instead of some boring communications guy back on Earth, knowing that Captain Braxton and crew were cruising the heavens above him. *Yeah, screw that*, he thought to himself. There had to be more than that to his mundane life.

"Wow, so a space fleet, eh? That's really ambitious of the captain. Wow." Tom chuckled out loud, more impressed and intrigued with the ideas being told.

"You don't know Kate Braxton until you're close to her inner circle, and that's kind of hard for most people, but you should talk to her about that if you're interested," Rachel reported.

Thomas sat there dreaming of the odds. He could be a captain, shadow the captain on his off time, learn the ship's layouts and all the stuff a captain would have to know. Or he could just be a pilot, or maybe switch trades and became an engineer. But then how would he go about that? All these ideas and thoughts running through his head made him start to get anxious and jumpy, and he had to get up and take a breather. "Hey, I'm just going to go get a coffee from the mess, maybe check out the celebrations and stretch my legs a bit. I shall be back shortly." He stood up and walked to the rear bulkhead, where the corridor to the lift was.

Thomas walked through the doorway and past another set of hallways on either side of him that led to a dead end with quarters on both sides. He guessed they were the captain's and commander's. He kept walking a few more feet and then came to the lift and stepped on, pulling the hydraulic lever to Deck One habitat. It took a few moments as the lift descended below and came to a stop with a click sound.

Deck One was fairly large; it looked larger than it really was, he thought as he stared at the large observation deck in front of him with its wide-open space and three huge, oval windows facing forward. In front of him stood the bulkhead walls; the mess hall was on the other side. Walking off the lift, Thomas came to the doors of the mess hall. They opened up, exposing a full room of ecstatic crew members having a good time. Smiling, Thomas thought, *Hey, this could be a good scene. Find some cute girls to hit on during this trip—this could be even better.* He laughed at this as he walked into the loud crowd. He spotted the chief engineer and the commander, then turned toward the little café counter where Cassidy was standing and having a good time. She saw Tom approaching with the same look in his eyes he always had when he bought a coffee from her university café.

"One special blend, please," Thomas ordered as he leaned up against the counter and watched Cassidy, who was already making his coffee. She told him it was coming right up. The maker dinged as she poured him a fresh cup and handed it to him. "Thank you," he replied, grabbing the cup and sipping the hot, yummy cup of Joe. It was just what the doctor had ordered indeed. Tom loved coffee, and he loved the Joe that Cassidy made especially. It was funny to know that behind the galley was the ship's hydroponics bay, where she was growing more of her special beans that made the crew's favourite brew.

"Are you all good, honey?" she asked him as he stood there dazed at the wonderful taste that was the ship's coffee. Tom looked at her and smiled then told her thank you again and walked slowly, disappearing into the crowd. Cassidy smiled as she watched her once students and now her fellow crew members on a spaceship they'd built, and a home they'd given her. She was distracted too in the back of her mind, as she was planning on making a maiden voyage celebratory meal for after they passed the moon during the test flight. It was going to be a yummy pasta dish with lots of cheese and hamburger and sauce, and maybe even yummy Caesar salad to go with it. The crew would love it, Cassidy thought, as the first days in space involved lots of work and probably strained nerves while systems were run and tested. Nothing makes a break or lunch like a delicious homemade meal by Cassidy Bishop. But little did she know that within a few weeks, she'd be cooking more comfort home cooking then she'd ever done before. She adjusted her stance and leaned forward on the counter, gazing out the three large fore windows. She lost herself to the classical music and the crowd until she dazed off into her own world. She relaxed and watched the room pass by as she slipped into a meditative day dream … and she was gone.

CHAPTER TWO

Mess Hall, 12:00 hours

The happy and cheery crowd was interrupted suddenly by multiple loud alarms going off simultaneously on the crew's EPads. The crew, taken by surprise, hesitantly reacted, slowly pulling the EPads out and mumbling and groaning. Then they noticed what was going on. *Valkyrie's* code red ship alarm was sounding with the typical wailing: *woooooop, wooooooooop, woooooooooop*! like a naval ships sounds before it goes into battle. Red strip lighting could be seen around the room. It would staystale like a red trffic light would when you pulled up to it at an intersection, and stay stale red until shut off. The partying crew now found themselves confused. As they were turning off their alarms, they saw the captain run out the mess door.

"Tell me this is a drill!" demanded Kate in a hidden, semi-worried tone. Only she and security knew the reason for code red while *Valkyrie* was berthed in the woods. Captain Kate Braxton stared into the screen of her Epad and saw her security guard. She talked to him through facial screen communication as they explained to her that the farm was going to have some unwanted company. Just down the farm road was a convoy of three cop cars and six black ghost trucks with red and blue lights. They were speeding silently up to the property gate. Then they reported that the front gate had just been smashed by the police, and they were only five minutes away from *Valkyrie.*

"Shoot, Captain, the police are coming. I can hold them off for five minutes, but from there, it is what it is."

Kate looked worried as a pit in her stomach suddenly sunk deeper. She knew that a police raid was possible, but now of all time? She breathed in deeply then exhaled and replied to the security guard, "Ok, delay them. I need five minutes tops, then go and scatter. But be on standby for further communications. Braxton out." Kate turned around and walked back in, cursing to herself and asking why now. So convenient this timing was!

The crews looked at her as she walked in holding her Epad and getting everyone's attention with her arms and shouting out for all to hear. "Can I have everyone's attention please? Thank you." The crowd shushed to hear the captain's explanation. "The farm is being raided by the police; they're on their way and will be here in less than five minutes. We all know what this means if law enforcement gets their hands on *Valkyrie,* and what happens to us, so what are we going to do? We have less than three minutes to decide."

The crowd broke into a surprised shock and frenzy over what to do as the crews talked amongst themselves for a few seconds. Large numbers of engineers sprinted out the mess doors, with Elisa Torres yelling, "Give us five minutes, Captain! We'll have the ship's engine up and running and ready for departure. Then she disappeared with her crew.

The other officers spoke out, dividing themselves into teams and also sprinting out the door, with orders and preparations heard down the hall. Commander Ahote and Kate were left there looking at each other in amazement at the crews' reactions, but they weren't surprised. Even Kate would be heartbroken to see this ship and everyone on it handed over the federal government. "Right then—to the bridge."

Cassidy was watching this episode play out with wide eyes of excitement. The captain and commander walked out the doors across the observation deck floors toward the lift as Kate explained that she was now more nervous about the law than the test flight. At least on the flight if anything went wrong it would be on their own terms and they'd know the results, instead of the law taking it all away and shushing it up. They got to the lift as the commander mentioned that it was funny that it took this long to catch on, which made him wonder if someone had rated them out. Kate didn't like this thought, and her face turned serious and went red with frustration. Her heart was pounding a hundred beats per second in her chest; the commander likely heard it before she broke out with, "No, no, no! We do not need this right now, but if we do have a cell member aboard, then we'd better find them before any more problems arise."

"Agreed. I'll get on it as soon as we enter orbit," stated the commander.

The lift made it to the bridge deck with a *click* as the two officers walked off and down the corridor, where they could see the windshield screen. But when they entered the bridge, they saw that all crew members were at the consoles plugging away as a low hum and vibration was felt through the floors.

"Aye, that's anti-gravity online and harmonizing," called out Kyle White.

The captain and commander took their seats as engineer crewman on console called out, "Captain, Torres is reporting that warp core is online and holding on standby so far. We're a go, Captain. Flight engineering checking in as go."

"Great news, crewman, thank you. Thomas, I need you to tap into the town's radio channel and monitor ground activity. I want to know what the law knows and what they plan to do about us. Keep me posted!" Tom responded by nodding his head at her, then he turned around and went to work on his console.

Helm called out, "Captain, helm is ready and standing by for lift-off orders."

Kate tapped a button on her small computer console on the armrest of the seat and monitored the screen with a list of systems and stations, watching each one go green, which meant ready for lift-off. Finally, in a few more moments *Valkyrie* would be ready to depart, but then came her Epad alarm. She quickly picked it out of her sleeve pocket and pushed the button to see the security team on the screen.

"Captain, we blockaded the barn yard entrance. Oh shi … run!" Then the screen went silent.

"What happened?" asked the commander with major concern on his face and in his tone as he sat forward in his seat, anxious as can be. The atmosphere was now chaotic as a happy and calm departure became a race for time as the systems in the science part of the ship were down to two systems on standby. Kate called out to the sciences stations on her Epad and ordered that harmonics and the anti-gravity be fully up and working—like right now!

"Mr. White, I want you to load the EMP scatter field and fire when I tell you to!"

He looked at the captain with a wide grin. "Aye, Captain. Ouch we gonna shut this place down something goooood."

The crews chuckled as she replied with a smile. "Oops, sorry Ma and Pa. I sure hope not."

"Finally, Captain, both port and starboard nacelles are charged and ready to go. Harmonics reporting all green, and the anti-grav rotors are at maximum levels and holding steady. We are ready for lift-off," reported Pilot Blackwood.

Kate looked at her screen and saw that all systems were green on her end. She looked up and commanded with a determined smile on her face and a desperate now-or-never-tone. "Do it now. Helm, take us up at maximum anti-grav velocity."

"Aye, captain," Rachel replied as the bridge interior lights dimmed and the hum raised to a slightly higher vibrational whine.

The atmosphere felt physically lighter in the bridge from the anti-gravity rotors in the belly of *Valkyrie.*

Just then, Commander Ahote called out, "Ummm Captain, it appears the cargo bay doors on the ops pad opened but now just closed."

He looked at the captain, who called out "Shoot! Engineering, do we have confirmation that the doors are closed and secure?" She looked at the crewman at console now with a frustrated glare. The engineer looked at the console for a few seconds and then looked up and reported a message from cargo bay. The ground security team had just been let in, but the door was closed and secure now. Kate just shuddered and said nothing as the *Valkyrie* revved her engines and the ship started to shake a little due to turbulence from the ground and the vibrations as the craft lifted up and away. In the bridge, Kate watched for a few moments, which felt like an hour, as the trees and plants hiding the vessel slowly fell below, bringing with them the tarp meshing covering the tree clearing. Moments later, it eventually snapped from its lines and slid off the ship, thus exposing the view of the whole town as the tree line fell below the leading edge of the craft.

Valkyrie was now hovering over the trees, and the sunny day shone through the windshield without a cloud in sight. A sunny, warm summer day like this provided perfect flight conditions, but then the bridge crew was faced with seeing their town as they rose above the whole community below. Then it disappeared from sight, leaving the mountains and the blue sky facing north.

Kate's energy was alive inside her body and stirring wildly with the amazement and wonder she felt, plus the anti-grav vibrations that were protecting the ship's crew from excessive speeds to be encountered soon. She started to tear up at the success they'd toiled over for two years. The trials they experienced, the ups and downs with the computer systems … she was proud while staring out the windshield. The reality of it all overwhelmed her.

The crews' reactions were similar to the ship's—vibrations that emanated a calming sensation while they concentrated hard on their tasks and their laughs of victory. Kate let them have their moment before she planned to bring the bridge back to their jobs. After a few minutes, she spoke up. "Ms. Blackwood, take us up full speed, ant-gravity dampeners on." The ship's pilot looked at the captain and nodded and then turned back to her console. Within moments, *Valkyrie* edged forwards and then up as the town below disappeared beneath them.

The ship climbed at a forty-five-degree angle with just light blue sky as the pilot read out the altitude. "Captain, we're climbing with very little turbulence, and dampeners are holding at 100 per cent, now two-zero-zero feet and climbing. Captain we're at one thousand feet ... twenty thousand ... thirty thousand ... now fifty thousand feet and climbing pretty fast now." The commander smiled as slight turbulence rocked the ship, and the sky was now turning a darker shade for a couple of minutes more before the turbulence stopped. Outside the windshield, the sky was black with little stars now showing up one by one. The moon was now clearer and brighter than ever. The ship's hum was now softer and really light, as a weightless feeling with gravity holding them down.

"Wow, that was a quick climb, Ms. Blackwood. Good job. Are we—?" asked Kate, stunned.

"Aye, Captain, don't blink. The ship's clock shows we were ascending at least five hundred knots and lower with some atmospheric deviations, but yes, we're in ... space." She turned around in her seat with a wide excited and satisfied smile, which was very contagious. Everyone looked at each other, clapping and cheering while congratulating each other.

Captain Kate Braxton stood up and walked to the pilot's console and stared out the front window at the blackness doted with stars and far away planetary systems, lit up with a few odd shaped, ovalish stars in the very faroff distance. She reached down and pushed the intercom button on the comm console and spoke. "Attention, crew. We did it! We made it into lower orbit and are currently heading to outer orbit, where we'll come to a complete stop and take in some scenery and do a systems update. Congrats to everyone. Your hard work has paid off, and we did it. Welcome to space. Ms. Blackwood, take us into outer orbit and come to full stop."

"Yes, ma'am," the pilot replied.

Kate turned to Kyle and ordered him to disengage the emp scatter field, and he complied. Commander Ahote called to Thomas Hilton, the ship's communications officer, for an update on Earth's chatter but got a stand-by answer. Then came the report of major chatter from the radio stations in the Fraser Valley, which were flooded with UFO sighting call-ins. Then law enforcement on the ground called in a cleanup crew immediately. There was nothing too special besides closing off the Braxton farm and securing the family for questioning.

Kate paged the chief engineer's work station screen console from the captain's chair Epad. Moments later, Elisa Torres appeared. "Hey there, Captain. I have initiated the system scanners and shut down the ship's engines. I'm happy to report that the warp core is holding. I'll keep you advised."

"Great work, Chief, thank you. Remind me that I must meet with you soon over coffee at Cassidy's."

Torres smiled and accepted her invite, but she had to get back to work, and with that they cut the connection. "You heard her, guys. Enjoy the view for twenty minutes."

Kate got up and decided to go to the mess to get a glass of champagne and hide in the corner and watch the view from the enormous oval observation windows. But then she thought, *Wait—the astrometrics lab!* It was perfect because it was located at the very top observation deck in the rear above the cargo bay, and no one would be there right now to question her choice to calm her nerves with a little spirits. She excused herself and walked to the rear bulkhead and down the corridor, past the private quarters hallways and straight to the lift, where she pulled the lever and was taken down through the opening. It came to rest on Deck One, where she got off and turned the corner. She continued to the forward section, passing a few science officers and saying hi to them.

The mess hall was empty when Kate walked in, and she noticed a half empty bottle of champagne sitting on the counter of the galley. It was funny because she must have been in the back or in the hydroponics bay and left it just sitting there. "Score!" she whispered out loud with a huge ecstatic smile on her face. She grabbed the bottle and casually walked out the doors with it under her arm. Kate then continued back to the lift when she remembered an issue needing to be dealt with: the security teams had snuck on board and were kind of … umm, stowaways! But then again, some extra hands might not be a bad idea, and the ship had quite a few rooms to accommodate these people. She chuckled at the thought, not really angry but annoyed that they now were an added expense to the ship's resources, but only by a small margin—not enough to be a threat to survival. But now she had to deviate from her plan and decided to go hide the bottle in her quarters and then track down the team and talk to them in the briefing room.

Five minutes later she walked down the access corridor to the right of the lift that led to the cargo operations bay. Just halfway down the corridor to her left was a small, empty office, big enough for ten people to have a console and a few racks and shelves with a small computer console in the middle with the ship's layout on display. Kate walked in and saw the five guards and motioned them to follow her as she led them back to the lift and up to the top bubble. Just around the corner to the right they entered a small room with a long wooden table in the middle and ten chairs on both sides, with one window that you could look out and actually see a little bit of the hull down to the leading edge. "Take a seat," she ordered as they all took their seats and she began. "So it's funny that I'm seeing you again, but then again, a few extra hands on board won't really hurt us any, so I see no issue with keeping you here with quarters and duty shifts, just like the crew." She breathed in and smiled before asking what they thought so far.

They were indeed speechless and not quite sure of what they were on or where they were. All they knew was they were on an airplane with a weird engine. "Follow me."

Kate smiled and got up and then led the team to the bridge, where their eyes lit up as they stopped in disbelief and amazement at the view of the Earth in front of them, blue and bright with white clouds hovering over oceans and their shadows reflected on the seas below. They were utterly unable to speak at that moment, until one of them stepped forward and stated, "This is God's creation—utterly magnificent, this unheralded beauty." His face lit up with sheer excitement, and he repeated his comment. Kate was going to explain the take-off and the ride up and give them a crash course on the C.S.S. *Valkyrie* when the quietness and human inspiring awe was shattered by the pilot's console alarm proximity sensor, which let out a loud *beeeeep, beeeeep, beeeeeep* high pitch sound.

Ms. Blackwood's eyes went all squirrelly as she was reading her console, which showed that something huge was right in front of them. Nothing was there until she looked up again to see a spatial distortion of some type forming a few yards off the port bow right in front of them. At first it was slow, but then it started to get bigger and bigger. It looked like a greyish cloud with a nasty tail wind approach. It took the form of a worm hole or a tunnel or portal of some kind, like a rift or something surrounded by multi-coloured lights spinning in an orbital pattern that got bigger and bigger before a fair-sized wormhole opened up. "Captain, I'm picking up an energy discharge on our port bow; it's like a worm hole, I think. Space is nothing like they told us."

Kyle reported that he too had the same readings and continued researching his sensors, which showed an anomalous surge of negative energy forming onto a vortex with an object approaching from within. "Ummmm, Captain, I'm going to say this once and clearly: there is another craft approaching our coordinates, and it's slowing down and coming from inside that vortex. It is indeed a vessel with a transponder." Everyone now looked up at Kyle with eyes wide and attentive faces. Their hearts were beating as fast as they could, and their minds were going into fight or flight mode. Energies stirred and breath turned into gasps at the readings coming in. The bridge wasn't blaring with external proximity alerts, and now the security team was also frozen is awe, but mostly shock.

Kate and the commander both turned to the windshield to see a large, odd looking vessel emerge from the star-gate-looking vortex that just vanished when it collapsed within itself like an implosion! The vessel was three football fields long and shaped like a long, narrow, three-dimensional delta triangle with the winglets or horizontal stabilizers like those on a Blackbird stealth bomber in the movies, only with a strut that held what looked like a vertical fin or stabilizer on the port and starboard struts. It was like looking at the lockhed Blackbird stealth aircraft from the front view, but it was thick and large with sharp and smooth edges that would be great for cutting through water and/or flying through space. It would definitely help with speed.

Jaws drop in disbelief and hearts pounded hard—no with fear or joy but just shock. Mr. White noticed this as "Wow!" was uttered in unison by the bridge crew. The object approached slowly, pitching its bow up a few degrees and exposing three large half bubbles sticking out of the bottom of the hull, obviously all bubbles at different ends of the ship. "Now what in the hell are those things? I'm guessing they're some kind of anti-gravity exterior plating or something," Kate thought out loud, still stunned and paralyzed by the enormous craft that had now stopped and held its position ahead of them, about two hundred feet away.

Kyle broke in again. "Yeppers, it's definitely another vessel. Scans show many life signs aboard, and many weapons systems I'm not familiar with. Her harmonics are off the charts. Yeah, Captain, I have to admit from what I know about this tech on board *Valkyrie*, it gets more complicated. Let's just say whoever they are, we should consider being careful with our actions."

"Yes, I'm getting those readings too, Captain," reported the pilot.

The crews' faces were glued to the large vessel. Ahote got onto his Epad and messaged Elisa, who answered back in a normal busy tone, obviously oblivious to what was happening outside. "Yes, Cap … ohh, Commander, h-h-how can I help you?" she asked as if caught in a chore, as she was just about finished with the system scans. "Ummm, give me five minutes and we should be all ready for the results."

"Elisa, come to the bridge please, and bring Pierre too."

"Yes, sir. Is everything ok?" she asked, now curious as to the commander's edgy tone. He replied to just hurry up and get to the bridge as soon as possible.

Just then, a loud screeching sound was heard and rang out throughout the bridge, thus making the crew frantically cover their ears in discomfort with their hands to block out the noise. In a few seconds it stopped. "What the hell?" Kate complained as she sat back into her chair.

A whirring noise seemed to come from all around them as a beam of light appeared in the middle of the bridge and a transparent human figure formed, wearing a well-fitted dark uniform of the military dress type. On the left chest were Navy and Air Force decals. He was a tall, Caucasian, Earth-looking human with brown, wavy hair and green eyes. His manly face was tight and serious. He had dark, sunken eyes with a heavy brow. He was very classical looking and well distinguished in his uniform, looking like a military leader of some state. Beside him stood a woman of Asian decent, very quiet and at attention in a woman's version of the same military dress uniform, with Navy and Airforce insignias. Kate had to admit that they knew how to dress themselves, but then again, it could be the uniform.

Just then, the security team burst into action as they quickly secured the bridge in a defensive and protective formation around the crew before the Black security guard spoke up. "Bridge secure, Captain, but if holodude does anything stupid, you can bet we're going act against it."

Kate and Ahote looked stunned but impressed at the display before the commander mentioned that there might now be a need for a ship's security team after all. Kate looked at him in agreement and nodded. She stood and walked up to the figure as he turned and looked at her. "Welcome to space captain Kate Braxton and crew," he said, "but I have to insist that you turn your ship around, Captain, and go back to Earth and surrender to the authorities. Do this and a brief interrogation will follow; after that, you'll be given state federal pardons for this escapade of yours. It's truly marvellous the brilliance you've shown, but now it's time to stop playing star explorer and head back. Besides, you have no idea what you're in for out here." He continued smiling handsomely while scanning their faces before he continued "I'd hate to see such young minds capable of what you've built be destroyed and wasted, when there are lots of occupations on Earth for your kind and like-minded individuals. Please land at the coordinates being sent to you, and all this trouble goes away."

Kyle tsk'd under his breath, yet his face didn't look as surprised as everyone else's, nor was he shocked, but he tried to keep it under wraps best he could and not blurt out what he knew. He was getting annoyed at biting his tongue over and over. There was so much he wished he could tell them. Ohhh, his heart felt heavy at the knowledge he didn't want to hold back. It wasn't because they couldn't be trusted, but why spoil this adventure for them? And there were certain things they had to find out and deal with on their own. As you reach a certain point in evolution, you transition to becoming space fearing, but now this was a whole new playing field he couldn't interfere with, and it saddened him. Kyle just put on a normal face and forced the thought from his head.

Kate looked up and replied, "You know my name? Who are you, and what are you of? You look human, and your vessel looks like a familiar Earth design."

The male figure laughed and replied, "Captain, Captain, this isn't a greet and meet. You are not to be here nor are you allowed to be up here with this tech of yours; now I order you to take our offer and go back home while you still can."

Rachel and Thomas were now getting critical of this character who was so bossy and overbearing; they could hear and see it as he tried to be nice and hide his superiority trip under warmth and kindness. He was very condescending, like toward small children, but they knew that this "kindness" wasn't going to last much longer. They looked at each other now, their eyes communicating that they knew what was going on and what was next. Thomas started an attempt to hack the vessel's comm systems as Rachel started to plot in a course to Pledan at warp 10 in case she had to act on a moment's notice. The security team was getting anxious, and their body language showed their distaste of this intruder telling them what to do after everything they'd accomplished.

Just then, Elisa and Pierre walked out of the corridor and onto the bridge, where they stopped and gasped suddenly "Captain, what's going on?" Elisa asked, stunned at the figure standing in the bridge and then seeing the vessel out in space. "Oh my God, what is that?" she asked, squinting her eyes and tiptoeing toward the banister, studying the vessel's features.

"Engineer Elisa Torres, I presume. Nice to meet you. I need you to convince your captain and crew to turn around and land and surrender and get a pardon for this. You're a very smart girl, and I know you helped build your ship; I wouldn't want someone like you getting hurt up here. You too, Mr. Gagnon."

Ms. Torres was now angry, and her face was getting red as she felt the pit of a heated attitude travelling up her body and into her head. Then she lost it. "How the hell do you know my name, and me, and him? And who the hell are you to tell me what to do?"

Kate turned to her and tried to hush her tone. Elisa got it and quieted herself. Kate spoke up. "Well then, you obviously know who we are, so you must know our attitudes regarding your demands. After that raid on my farm, I have no intention of paying kindness to a malicious act against my family. Your offer is nice and all, but it seems more like a threat with a dagger at ready behind your back. And as far as not knowing what we're getting ourselves into … well, let's just say it's not up to you to decide our paths, so buck off, pal." She breathed in, her face all red, and then continued. "You can tell us what to do all you want, but this is not your adventure or your path, and you certainly didn't build this ship from nothing or you would know that this conversation means nothing to my crew. I politely ask you to back off and forget this ever happened. You don't even know why we're here and what it took to get here, so I'm asking you to shove off, pops."

The bridge crew snickered quietly under their breath, now seeing a part of Kate they'd never seen before. They were impressed at her will to not be walked all over, and this brought some comfort, knowing that the captain may certainly have some gumph to lead them on this voyage. It was true that Kate did not like being talked down to, or being treated like a child and told what to do by authority who thought its crap didn't stink. And she wasn't keen to hand over *Valkyrie* at anyone's request. No one on the crew would willingly hand over the ship to anyone, let alone some government agency that's not to supposed to exist and nobody knows about.

Out of the blue, Elisa charged up to the image of the woman and demanded in a commanding tone "By the way, who the hell are you, and how do you exist and no one knows about this?"

Kyle pushed a few buttons on his console and texted a message to the captain's chair, where the message indicator beeped a few times. Ahote tapped the screen, transferring the message to his screen and read it: Captain, I would suggest at this point that we break free and make a run for it ASAP. Helm is ready to warp out, comms is hacking for chatter, and I've armed the EMP scatter field, which they can probably see right now. A getaway would be suggested, like right now. White." The commander looked up at Kyle, who was already looking in his direction, and nodded at him. Ahote looked back at Kate and nodded her to her chair as he stood up to replace her at attention, ready to act. Kate sat in her chair to read the message and then looked up to glare at White. She caught his attention and nodded, because deep down she knew this conversation was going nowhere real quick.

The figure now looked angry and responded to Kate's comment. "You little brat! You'd be best to heed my warnings; you will never get away. Deny me and you can just rest assured you'll never get to your destination. Last warning, Captain!"

Kate turned around to the male with a tone of conviction and replied, "No. No, we're not surrendering. Good day, Admiral whatever. Kyle, do it. Ms. Blackwood, execute." Kyle nodded and aye aye'd Kate as a small shudder shook the ship and a blue colour, full light wave in ring formations was released and headed for the strange vessel. It too shook, leaving the fore half completely darkened out, with the midship windows flickering on and off while the rest of the ship had lights. The male figure now flickered a few times, glitched out, then disappeared.

The pilot was busy at her console as *Valkyrie* pitched to the starboard and sped up engines immediately. "Aye, Captain, course laid in for Pleiades and warping out at warp 9 … now!" In a flash, the C.S.S. *Valkyrie* was gone, warped out and nowhere to be seen, leaving the strange military vessel partially damaged and shut off, drifting with systems failing left, right, and centre.

Day 2, 0700 hours

Kate woke up slowly as she turned onto her back and stared up at the open skylight window at the head of her bed. She always loved window seats on planes and liked having the stars staring back down on her. It was weird, but she wasn't unnerved at the fact that space is hostile, and the windows that loomed over her provided a pretty good view of space outside. The vessel's antigravity rotors created a calming hum that resonated through the ship and made the quarters quite comfy. The quarters were all the same, just a basic medium-size cabin with a nice washroom and a small divider hiding the beds for privacy. Then there was the main area and the quarters' door, but the furniture was very nice and modern looking. The lighting was the same, but with colour mood lighting with blue and purple options and shades. *Valkyrie's* quarters were indeed very nice accommodations and facilities for the crew and a few visitors.

She finally decided to get out of bed and go get a cup of coffee before heading to the bridge. The crew had been quiet lately, still digesting the fact that other humans were out here also, and the encounter above Earth with that vessel. Some crew still wondered how this could be, while others felt happier and a part of something bigger as the first civilian Earth humans in space. A few were still coming to terms with what this meant and what could come of this discovery. But all crew are well aware that this was a game-changer and had probably changed human events in the solar system. It still amazed Kate that her crew were such deep thinkers and knew that *Valkyrie* being out there had already made ripples in history. Probably everywhere in their galaxy could be affected by it, so the way they acted and the choices they made would greatly effect others. Eventually this meant their human flawed attitudes and tribalisms were carried with them. But then again, they were a crew of selective people with the right attitudes and mindsets. Kate knew they would find common ground on their interests, and when two interests or people met, those bridges would be quickly rebuilt to connect them. That was going on right now. The crew was good, and Captain Braxton had full-on faith in them and their new existence out in space. But what was spooky was the fact that they could beam a hologram of someone into your ship, and that downright disturbed some of the crew. Others weren't surprised one bit and kind of expected weird and funky things to manifest from time to time, just like in their favourite space shows.

Getting out of bed, she pulled herself up, grabbed her uniform, and put it on slowly. She was still thinking about the encounter and the two university students who discovered the Pleiadean lights, and how all that had led to this moment. Now it appeared that other places must be teeming with life, and space truly was turning out to be the final frontier. Kate put on her uniform and touched up, sliding a hairband over her head. She walked out of her cabin to the right and then to the left, boarding the lift. Walking out of it, she continued on her way to Cassidy's.

Kate passed by Mr. Gagnon on his way to engineering. She gave him a smile and a "good morning." He reciprocated the greeting with a truly happy smile like she'd never seen before. She kept going and walked into the mess, where Cassidy was serving an egg breakfast with green and red peppers and mushrooms sautéed and fried in butter. There was also orange juice or coffee. "Morning, Captain. Come, child, get a good day's breakfast before bridge duty," she called out.

Kate smiled and walked up to the counter with everyone else until she was served a hot, fresh pot of coffee and an omelette to die for, with ham and pineapple and bacon and three different cheeses. It was delicious. She sat at the table with Elisa and a few other engineers on her shift while indulging in her early breakfast break. Engineering was pretty quiet at this time according to the crew reports; all systems were functioning normal. It was funny that Kate saw space travel as easy. How it worked was the hard part for most people to comprehend, but she had a lot in common with the engineers when it came to making theories possibilities. They talked about the possibility of discovering a transwarp past warp 10 by amplifying the extra negative energy with a stronger warp core, and maybe redone transducers. They could send in a probe for a trial run, which the engineers and science teams could build and launch. It was doable, they thought, as a smart system could be built that was connected to a remote control with an amplified signal harmonics system that could bounce long distance signals across space at a frequency that would bounce them off a source that strengthens the signal and bounces it back to the coordinates it came from. This was an exciting idea because in space it seemed that everyone was in higher and positive spirits and felt a lot looser. The crew also reported being more creative and enjoying a higher intellect. It was a refreshing discovery, and Elisa and a few others noticed it too; indeed, the *Valkyrie* was alive with a vibrant and intelligent crew that moulded the positive elements into one perfect family.

Everyone was happier and brighter; they were doing this not because they had to but because it was a part of exploring, and it gave them something to do and supported their chances of exploring longer. Little did the crews know that the next one hundred days were going to be very smooth and quiet. New members became crew dedicated to defending the *Valkyrie* and its crew. Engineering also planned to upgrade its ship-to-satellite systems and give a boost for message sending, which would be a big help for crew letters home whenever an Earth satellite was in range. Even the pilots con was to be installed with a five-minute warning system and alarm to alert the crew of possible contact to loved ones back home, using the satellite system as a personal email account. It was a great time to be a crew member onboard the *Valkyrie* heading out into the unknown to find the answers we all seek, but be aware that some things are not set in stone and determined, nor are they what they seem to be.

CHAPTER THREE

Kate Braxton, Deck One

Kate walked down the main corridor with a cup of coffee in one hand and her Epad in the other, reading the crew reports on her way to the unused office where the security team was waiting to meet her for a briefing and proposal. She was pretty unnerved by the encounter with the Earth humans in that huge ship, but she had expected to see other life out here, just no so quickly and close to home above Earth's atmosphere. It was a bit of a surprise for her and everyone else on the bridge, mess hall, and the observation decks. Still, she had a deep, deep feeling in her chest, a stirring if you will, that there was life everywhere out there. As scary as that could seem, it was quite exciting to think that at anytime, they could come across someone else. Kate turned left around the corner and then left again, past the lift and the virtual reality room, until she turned into the hidden corridor that turned to the right. She walked straight ahead for twenty feet into the tucked, out-of-the-way offices that were meant for other uses and stood empty at this time. These offices were on the other side of the operations and cargo bay, where the security team stood around an unlit console table talking amongst each other.

"Ok, everyone, that was a well displayed setup on the bridge. This changes everything now." Kate went on to explain that as a civilian vessel, the crew were welcome to stay with no punishment for stowing away, but there was the issue of food and supplies, so they would have to pitch in on duties in the hydroponics bay to set up more planters for more crops to grow and help tend to the few chickens and goats they had on board. They would also help to grow lots of peanuts and soy plants for meat alternatives the crew would consume with their meals for protein and energy. Since they now knew they weren't alone out there, they might as well have a security team ready to deal with ship boarding and protection, and maybe to guard the bridge from time to time.

Crewman Morrison stood up in a military stance, with his arms behind his back, and nodded, thanking Kate for her choice to keep them on board as active crew members and just to be included in this awesome and amazing endeavour that would soon change their lives more than they could ever know.

"Well, now that you've been briefed, let's talk about your department and your tactics to keep this ship and her crew safe. Any ideas, anyone?" Kate kept shifting eye contact with all security members who acknowledged her. They fell silent for a few seconds before one young woman spoke up.

"We need weapons or something, like guns, in case we're boarded, just like that hologram the human ship beamed onto our bridge."

A male also piped up. "Yeah, I have to agree, Captain. They were able to easily send a hologram of their captain's image and beam it onto our ship. Who's to say we won't run into other races out here that can do the same thing, or even more? What if next time it's a hostile race, or a thief, or no-gooders? Where there's good, bad is always soon to follow. This happens everywhere in life. What do we do then?"

Mr. Morrison turned to the guy who spoke and nodded his head. "Good point." Then he turned to the captain and explained that they'd need this office and uniforms and weapons. Then they'd have to set up a crew roster and shifts on the bridge.

Kate was impressed at the ideas flying around the room and looked at her Epad to see that engineering was running a data scan on them at that moment from the engineer's consoles that lined the walls in the systems corridor running the length of the ship on Deck One. Morrison spoke up and announced that he was going to assume role of chief security if no one else minded, and he'd be making his first assignment with the engineers for rifle lockers that should be in the security offices and guarded all the time. The woman spoke up again and introduced herself as Sherry Wiles and requested a uniform for everyone and then some quarters too. Kate smiled, happily sipping her coffee as the security crew organized themselves. "Right, uniforms," she replied. "Go to the storage room on this deck and there should be one hundred extra outfits in there you can have, and an Epad each please. By midnight tonight I'd like a situation report on your progress. Until then, I must go make my rounds and see to *Valkyrie*'s smooth running." With that, she nodded good day to the new crewmen and turned around and walked back down the corridor, leaving them to their duties while she continued around the corner and into the main corridor and walked toward the bow of the ship, where the bridge was.

Ahote slowly woke up an hour later than the captain, as was his shift schedule, to the sound of his clock buzzing on his bedside table like an annoying bug that just keeps flying around your ear and never goes away. He'd chosen this alarm because it drove him to get out of bed eventually to shut the damn thing off, like swatting at it. He rolled over sluggishly and smirked out loud. He then looked up to his window over his head and saw a nice view of the heavens. They looked to him like they once did to his ancestors so long ago. Ahote took this in and stirred his head into a meditative state, but only for a few seconds, hoping to catch a little awe like his people felt back in the day. But the alarm clock was still buzzing, so he snapped out of it and announced out loud, "Right. Even in space, I need time to meditate." He pulled himself out of his bed and hit the top of his clock. The alarm shut off with a squelch that made him chuckle again with an even bigger smile. He was somewhat proud of himself.

Ahote walked to his closet and pulled out a second uniform and his native shawl, which he used for ceremonies and stories gathered around a campfire with family and friends and tribal groups. He figured that this moment he was going to embark on was equally important for him to wear it. Within a few hurried moments, he was out the door as it closed behind him. He snuck up the corridor away from the bridge entrance just behind him, thirty feet away up to the first corridor. He turned to the right and saw Mr. Gagnon heading for the lift as well.

"Morning, Commander. Interesting shawl you're wearing," he said in a confused French-Canadian accent, not knowing what was going on with the commander. Then he smiled as Ahote caught up to the lift and got on, smiling back with a nod.

"Morning, Mr. Gagnon. Just staring your day?" the commander asked as the engineer pushed the hydraulic lever up to the observation deck.

Pierre replied with a clam but distracted tone. "Yeppers, first is coffee, then I'm working on a request from the security team, something about rifles and lasers or other ways to make weapons for ship defence."

The commander's eyes went wide with intrigue for a couple of minutes, not knowing what to say. As far-fetched as it sounded to him, the idea wasn't to be dismissed. Just then the lift came to a stop at the observation deck. "Well, think of it this way. You guys got this ship built and flying through space. Good luck with this endeavour. I'm fully confident you'll pull it off." With that, Ahote smiled, patted Pierre's back, and stepped out of the lift and into the corridor, where he turned right and walked down a hallway.

He eventually came into the large open area with three thirty-foot large oval windows providing one hell of a view of space. They were facing forward bow, allowing people to gaze out ahead of them. The bulkhead walls were the same carpet and colour. To his right along another wall was a huge, lighted circular platform with florescent blue lighting and a holographic image of the *Valkyrie* flying away from the Earth and then floating motionless. There was a copper or brass plate at the base with words etched into it saying: "Dedicated to the brave men and women of the University of Chilliwack and now the crew of the *Valkyrie* circa. Thursday June 22, 2019." It was a monument made as a gift by one of the crew who was a designer of holographic art and media in her spare time. She thought it would make a great piece for all to remember and admire for their accomplishments; this artwork had caught his attention and made him think back to the day he was called to Kate's Christmas party a year ago, when she recruited him. Now he breathed in and out slowly, remembering that they had achieved a goal that seemed so impossible back then.

Ahote took a few minutes to focus on the moment, take in the feeling of the ship's soft hum, and connect to the moment and how he got there. He shed a tear and tightened his smile at the thought of seeing the crews' attitudes change and groups befriend and get closer to each other from day one up to now. Friends were made over time, and even stubborn individuals came together as crew members and learned to put their stuff aside and unite to build the *Valkyrie* and come this far. Any petty squabbles or smart ass remarks in the past now meant nothing and no longer existed, at least not that he could see. How soon the crew came together was amazing, and this made the commander kneel down and place his hand on the epitaph for a few minutes and breathe in and out. He reminded himself of why he'd come up here, which was to connect in a meditation session by himself before his shift started. Ahote found a quiet place in the centre of the room with the sound of pots and pans coming from the mess hall ahead of him. He laid a multi-coloured, dark shaded, orange and black and yellow striped mat down on the floor and sat on it crossed-legged. He then held his palms on his knees and began to breathe slowly in and out while facing the three large windows and staring into space. He called out to the heavens and whoever was there in his head while nodding off into a relaxing state and then into a trance. He now slowly connected with the view, cutting out the walls that divided the windows and allowing him to see one huge window with a full open image of the cosmic plain in front of them, just like the windshield view on the bridge but bigger. His attention was fixed on a bright point way out there that stood out; slowly he felt his energy inside gravitate to the star he'd fixed his gaze upon. With letting go of the moment, something truly natural feeling but creepy happened to him. His energy stirred and he just blanked out.

Elisa Torres sat at the control panel on her console across from the warp core that was shielded by two five-foot-thick transparent metallic alloy walls that shielded the crew from the core and any of its heat that it exuded while in warp. The core was walled in a chamber with a cooling temperature to keep it from expanding and warming up the engineering room, which made Torres at ease and happy, because she really did hate the summer and preferred the average room temperature the ship was set on. It was even comfortable with the other crewmates, but the chief engineer was busy running cool temperature tests on the core's overheating control systems while the other engineers maintained the core levels and updated new software programs to the ship's computers. Torres wanted to keep on top of maintaining the cooler system so she would have to face a breakdown, thus causing steam buildup and condensation on the transparent walls. This would block the view of the core and over time likely prevent an overheating of the core and its fuel contents. If that happened, she'd just bring *Valkyrie* out of warp for twenty-four hours to let the engine cool down before resuming at a lower warp speed.

Elisa had made sure that when she and Pierre built the warp core and the engineering room, they also included a few easy failsafes that would immediately cool down the engines and allow the ship to slow speed and recharge a bit and rest up its energy intake. The cooling chamber around the core was one of a few systems they had in mind when building the engineering room, thus making a few other easy and safe protocols that were gentle and effective on the engine with no risk of damage or catastrophic outcomes. This sometimes made her question if the sciences of space travel were actually easier than most think once you got the technology for it down and assembled.

The door around the corner from Torres to the engineering room opened with a motion detector. She heard a whining sound as it closed behind Pierre, who appeared from the corner with a smile and a good morning with two cups of coffee in his hand. She looked up and smiled, grabbing a cup out of his hand like a coffee addicted caffeine freak. She took a sip and then stated with a satisfied look and exhaling calmly, "Mmmmmmmm, coffee first thing in the morning, and it's Cassidy's brew too." She turned to him and smiled and then thanked him before transferring her data to his Epad, which beeped to alert him to a message or report being sent. Pulling it out of his pocket and staring at the details, he walked over to the wall casing console and punched in some numbers he'd run through his head the night before. He saw the algorithms match up with his previous calculations. Satisfied, he smiled and pushed a few more buttons and stated that he would log this report. In the meantime, he was going to his office to research security's ideas for rifles or pistol building.

Torres chuckled as he disappeared behind a corner with a computer station and quietly plugged away reading his request. In truth, all ship's systems and sub systems were running smoothly for now, and there wasn't even one indicator level that looked out of line or caused any concern, which made her bored and a little annoyed. But that's a good thing out there, as no stress or anything life-threatening was ailing her or the crew. This meant success, and this was what she wanted. But she didn't want to be glued to one console for half a day, staring at peak efficient levels that wouldn't drop at any moment, even though they could. Torres took another sip of her coffee, followed by another blissful taste, when the thought occurred to her to go to the mess hall and bug Cassidy for a while. With a snicker, she got up and walked out, announcing to the room and its four engineers on duty that she'd be back later and to call her if she was needed. With that, she walked out the door and down the corridor to the lift.

Staring at the floors pass by as she rode up to the main deck, she saw crew walking past and smiling and greeting her good morning in their smart-looking uniforms. She took another sip with a proud and smug face, admitting to herself that she was indeed the happiest she had ever been, just like the other crew were. Then she thought about her family back home and what they would be thinking, knowing that she was on a list with several other missing students who were on *Valkyrie* with her. She thought of the fact that when they launched that day, the captain had set a flight plan that took them low over the whole town of Chilliwack for everyone to see and be a part of. This made her giggle to herself, as she knew why Captain Braxton had charted such a flight path. Now she tried to imagine what their families were going through after putting the pieces together. Word would have spread from Chilliwack into neighbouring towns and cities like wildfire, and there was a pretty big chance that some people got the take-off and departure on their cell phones and downloaded the videos onto any and all social media outlets, thus leaving the ship and its missing students in history books and in the minds of those who believed and disbelieved that day. Then that lady, Marie May, came to mind, and how wrong she was. Now she'd be faced by the whole town questioning her and Dick's authority and church groups. That thought made her laugh even harder, thinking those religious people must be going insane. They were a whole other story at the beginning of this adventure.

The lift came to a stop as she got off and slowly walked through the observation room toward the mess hall that was separated with a bulkhead wall with pictures framed with the crew of the Valkyrie, and a lone figure sat in the middle of the floor, facing the front observation windows quiet and still. Torres walked toward the mess, glancing at the figure, whom she recognized as Ahote. He looked pretty out of it in a meditative stance when it hit her to just leave him alone. She knew he liked to do what he called astral travel sometimes. She passed him and went into the opening doors of the mess hall. Cassidy stood at the huge window in front of the tables and was staring out into the void of space. She too caught the image of a purple blueish gas cloud. She almost wanted to call it a nebula but thought it might just be a simple gas cloud. It was so beautiful and mesmerizing that she found herself drawn to the observation window to stare at the slowly approaching cloud, so bright and expansive yet light years away. It was truly a spectacle that made her feel alien yet intrigued. "Wooowww, you don't see this every day, or ever. It's soooo beautiful," Cassidy whispered loudly and calmy, as if she were in a spell from the cloud.

Elisa smiled and agreed with her eyes locked on the colourful cloud with different spectrums of dark to light-bland purple colors that hypnotized you if you looked at it long enough. Just then the cook turned to the engineer and asked, "What can I get you, my dear?"

Elisa took a sip from her mug and slowly turned to Cassidy and replied, "All systems are good and nothing needs doing at this moment. I'm bored and need to just kill an hour or so." The two turned to sit at the table behind them. She sat across from Cassidy, who also had a cup in her hands and asked what was on little Ms. Torres' mind.

Three hours later, Cassidy and Elisa were still laughing hard in their seats over their fifth cup of coffee. Now the engineer was feeling hyper but glad she was having a good time with the school's café owner, who would always nurture the students every break they got. But this was different, because Ms. Bishop was telling about when Kate had given her a strange note with Kate's address and a date a month later to show up on the Braxton farm, where she was introduced in the woods to the hidden makeshift hanger and a half-built ship frame. Students were working on open and uncovered decks, installing computers and consoles with a few hydraulic lifts carrying supplies and students to the upper decks not yet hull plated. She recalled her experience seeing the ship half-framed and the students she knew and served every morning working in this huge structure, which she soon came to realize was a spaceship. Heading this project was Kate Braxton, whom Cassidy suspected would attempt this idea after the plans were leaked. That had made the two burst out laughing. They enjoyed the spirit of the students as they secretly built the ship and studied in school the next day as if nothing was going on. The outside protests were getting louder and directed at the students.

It was amazing to Cassidy that no matter the naysaying that went on, the crew never once gave up or turned sour. They just seemed distracted and focused on something that wasn't school, because they were more joyful in their daily doings. Bishop couldn't tell what it was until she met the *Valkyrie*. And now Cassidy Bishop, the university's café owner and mother to all students, was right here right on this ship with them. She chose to be here as a vacation and time off from the school and the religious anti-first-contact protests and crowds harassing her after school. She wanted a break and took vacation leave to join the crew. Elisa was overjoyed to hear that Cassidy was also happy, and she noticed the changes in her mood and the crews', as if all their problems didn't exist. Many reported to the ship's doctor the feeling of being free, which panicked a few crewmen who were once overstressed and burdened with classes and books and exams, just to one day hopefully get to where they were now. This made the two girls laugh even more as they slapped the table and congratulated the students.

Just then, Elisa's Epad hailing alarm went off. She grabbed it and turned on the screen to see a science officer from astrometrics greet her and request that Torres come up to astrometrics immediately, as she was going to want to see something. She got up and excused herself and quickly walked out of the mess into the observation deck, where a few crew were now standing and looking out the window. Then she walked across the room into anther corridor that led another fifty feet to a bulkhead wall, where she turned right down a dimly lit hallway and into a fair size room with oval windows all around her. In the room sat a round table with a three dimensional star chart on it, with *Valkyrie's* path from Earth marked with a red line from the Earth's solar system to their position now. It was explained that the astrometrics lab created a scan of the surrounding areas for a mile in all direction and created a chart of what the scans picked up, kind of like the sonar subs that found and mapped the floor when the *Titanic* was found. By using this idea, they could collect information and create an image of the space around them up to a mile away. It was truly amazing to Elisa that this technology existed, and it was leaked to them by Kyle's contacts, who helped out in more ways than asked.

The room was also dim, with blue lighting and consoles running along the wall with mapping stations and information consoles for charting and scanning space. But this wasn't why she was being called to this room. Mariane Perez, a science student in astrophysics and stellar cartography, was standing at the middle console with another science officer, who looked as if she had seen a ghost. Elisa approached the round console in the middle of the room and asked what was up. Perez turned and briefed her, showing the latest reading from just five minutes ago from the ship's sensors just half a mile to the starboard side of *Valkyrie*. A subspace disturbance had shown up quickly. Its frequency lowered for a few seconds and then rose again and disappeared. Being curious about the phenomenon, and as scientists, she and her helper had run a sensor match with any occurrences the ship's sensors may have picked up since they left Earth. Then a match had shown up on *Valkyrie's* computers, and further research showed that the match was the same frequency as the unidentified Earth human space vessel that had shown up above Earth's orbit, except the frequency was of some energy readings and not necessarily a ship's readings. It was probably a wormhole activation and deactivation with something exiting entering another one before disappearing off sensors altogether.

"In short, something out there popped out of a likely wormhole and closed it while opening another wormhole and disappearing into it, meaning we might have a documented UFO sighting, Ms. Torres."

Her concerned-looking colleague looked up and spoke. "This means that the frequency matches the wormhole that Earth spaceship came out of. This has to be another wormhole reading, which means we're not alone out here maybe." Elisa now stood at the console, staring at the star map image of the event as it showed a pulsating flash of light and an energy stream fly out and come to a complete stop. The event closed and another event opened up in front of the energy stream as it sped off into the event and disappeared altogether, which left them speechless and frozen with eyes locked on the screen in disbelief at what the telescopic camera above them had recorded. Then Elisa spoke in an urgent tone. "Somebody get the captain now; she'll want to know about this."

The crewman nodded and paged the captain from her console just along the wall behind Torres while Perez and Elisa studied the video feed. Both agreed that this phenomenon had to be a wormhole closing and another one opening, and even that event frequency matched the strange military ship that had popped out of earlier. Perez had suggested that maybe it was faster than light speed or a drop out of warp or quantum drive of some sort. All they had was a bunch of theories that were backed by an event that had already happened matched up to the encounters.

"The captains is coming right now. I'm going to save all this to the ship's memory banks and record it for my report," said the science crewman as she grabbed an Epad and headed out the doors to the computer room in the bowels of the ship, leaving the two other girls waiting for Braxton's arrival. Surprisingly, in a few minutes Kate walked in with her second cup of coffee and a perplexed and inebriated look on her face. Torres smiled and watched the captain walk up to the console and with a hand signal told crewman Perez to report the finding. The astrometrics officer caught on and began to explain the reading they got off the ship's star mapping sensors. This made Kate's eyes open wide with shock, even though the captain really wasn't that shocked or surprised but more interested in these little things that kept happening every now and then.

Elisa chimed in and excitedly pressed a few buttons on the console keyboard to bring up the video feed of the event. Braxton was all ears and attention as she watched the screen for a few minutes, like she was deep in very excited thought over what to do next, but really Elisa knew the captain was going to just share this info with the crew and order the astrometrics crew to keep at their assignments and report back to the bridge immediately if anything else showed itself. Kate turned to Elisa and ordered her to report this. As she downloaded the information on her own Epad, she thanked Perez for the update, explaining that all they could do was save it to the ship's systems and keep an eye out. They might come across something they could alter course for and go investigate; then she walked out and back to the bridge.

Kate sat in her chair on the bridge next to Commander Ahote, who was also on his Epad studying the report with enthusiasm and a smile on his face. Kate glanced over with excitement in her eyes and spoke. "Pretty neat, huh? A possible wormhole activation or warp drive energy signatures, but I find it really interesting that the object that comes out and goes back into the phenomenon actually stops as the one closes and opens. Then it goes in and disappears like intelligent control."

Ahote looked back at her with a chuckle and stated, "I don't doubt that we've gotten attention out here. Think of it—humans as we know them haven't been up here yet, but here we are, the first that we know of. If I were them, I'd be curious too as to who we are and what we want."

The captain smiled in agreement. *Beeeep! Beeeep! Beeeep! Beeep! Beeeeep!* The sensor proximity alarm sounded from Rachael's pilot's console and let them know that an object was in front of them and closing in. The bridge crew suddenly looked up at Ms. Blackwood as she sat in her seat plugging away at her console while reporting that they had contact up ahead one mile. It was about the size of *Valkyrie* and was stopped and stationary. "Permission to take us out of warp, Captain?" the pilot asked. Then as suddenly as it showed up it just disappeared off her screen and sensors. "Wait. Captain, it's gone. Like, it's gone. Just disappeared. But I have traces of an … ion trail?"

Kyle reported that his readings also supported an ion trail leading off in the direction of the Alpha Centauri system a few thousand light years from them, but this signature had traces of life emanating from the object, although very little. The bridge went quiet with few hushed whispers while Kate stood up and marched over to the pilot's console to check to see if there was a match with any signatures seen by astrometrics and the human space vessel above Earth on the day of there launch. In a few seconds, a match was found in the frequencies and ion readings. When she ran a diagnostic scan on the particle readings, she was prepared to see what she had just discovered.

"Mr. White, I need your expertise over here," Kate said. Hearing that, he immediately got up and walked down to Kate and peered over the screen, only to confirm that the signatures all held a differential level of the same particles that were released into the trails, with traces of combustion and a few other isotopes known to be thrown into heleron colliders and other such combustion engines. This also showed that they were used for a combustion model for an engine drive of some sort, meaning that these ion trails were left there by an event caused by intelligent beings of some sort. "Oh dear, we're not alone out here, are we, Kyle?" Kate whispered in awe.

Rachael then suggested sending out a sub-space message saying hello to help curious ships identify *Valkyrie* as not being a threat. Ahote stood up and agreed with enthusiasm, which made the crew giggle a little bit with all the overwhelming emotions and reading s going off all around them. "We may find ourselves, Captain, in a first contact situation, so I think we need to call a ship-wide meeting and prepare for a possible contact with star people."

Kate looked back and nodded her head "You're right, Commander, we all need to be at our best. We need to be fair and open to whatever we come across, different or not. We need to set a new standard for ourselves, because this is a new final frontier. I have a feeling this is not yet the beginning of our adventure." Kate walked off the bridge and down the corridor toward the lift with sheer exuberance at the newfound information. Right now she wanted so much to put it all together and send it off to all crew Epads, but she had to prepare files for them first. Right now she needed another coffee, then she'd start and share the news with Cassidy, who was going to be very … well, Cassidy would be excited and a little emotional, but hey, she'd be pleased to see the crew's hard work and learning pay off in more ways than their families or professors ever thought. Then she'd laugh heartily, knowing that her crew were smarter than anyone imagined, and she did love them for it. Then she thought that Cassidy could be included in this face-to-face with other beings, as they would have to throw a banquet or sit-down of some sort, and the only place for that would be the mess hall, if the contacts were to be invited.

Before she knew it, she was walking through the door of the mess, with a new sign hanging over the awning, just like in the university. Obviously Cassidy had done some decorating. The new name of the café read: "Cassidy's Intergalactic Café." There she was with some decorations and banners in one corner set up to decorate the bulkhead behind her café and galley. It looked like she was going for a galactic theme, with a little bit of the corner where her university café stood, so it was like a little home with a space theme evenly balanced into a then-and-now montage. But who knew until she'd be finished with it?

Cassidy looked up and laughed while waving Kate forward. She had wallpapered with a picture of the university's foyer wall, identical to the real setting back home. Then Kate noticed that it had some dimensions to it from some angles she was shifting right now. It dawned on her that this was a three-dimensional hologram effect wallpaper, a pretty awesome idea for the crew, as it totally reminded her of the university and the comfort of the university lounge and vibrancy. It was the balance that the mess hall needed.

"Ms. Bishop, I have some exciting news for the crew, and you are the first to know because I need you to prepare for the event of first contact. Don't laugh until you read this report."

Cassidy's face went to a small grin and she grabbed the Epad from her captain and glanced it over, motioning with her hand for Kate to follow her to the couch lounge area in the opposite corner. There she started to read in great detail all the events that had unfolded and the sensor matches with other phenomenon. She looked up and whispered, "You're not kidding, are you, Ms. Braxton? Holy mothers and sons!"

Kate shook her head and smiled and then replied, "Now you are an ambassador of the human race if this happens. Are you up for it?"

CHAPTER FOUR

Deep Space, 24:00 hours, 100 days later

Rachael Blackwood sat at her pilot's console with another auxiliary communications crewman next to her; he wasn't very talkative tonight, though. On the bridge it was quiet, except for the low pitch humming sound of the ship's engines, soothing and relaxing as they were, and a few systems notifications alerts from the consoles. Present were crewman Emily Wang on science console, Pierre on engineering, and a few security officers on duty. Security was now a department and had been given the empty office. They'd turned it into a security office with an active ship mapping system with tactical access to a few systems on *Valkyrie* plus a storeroom they were going to turn into a weapons locker with the leftover diuranium scrap sheets from *Valkyrie*'s construction. They planned on making rifles and pistols with this. And let's be honest, ever since that run-in above Earth, who wouldn't want a security team in case they came across more ships like that. They set up twelve-hour shift operations that rotated every week to give all officers a chance to have days and nights off. They immediately went to engineering labs to talk about making some phaser firearms for security's twenty-four-hour Delta Omega Force, which was the ship's tactical response and escort division. They wore the same uniforms as everyone else. Security crewmen

Dan Morrison and Sherry Wiles sat in seats along the bulkhead next to the outer consoles.

The security team brought more comfort to the crews still struggling with their new reality, which had forced the ship's doctor to take up a new role of counsellor unofficially to deal with the anxieties of some crewmembers. But having security had made a difference in crew performance and mental health. Crewman Blackwood was running some scanner tests and looking at random coordinates that they had passed or were coming up to. At one point she had scanned some small Y-class system that seemed to have a copper element on the planet below, which was hazardous to humans if they didn't have elemental suits, which at this point they did not. But it was on a list of things to look into. Then there were readings of small ice balls floating about, but not catchable at the warp-six-speed they were cruising at.

Pierre was in charge of the bridge and was at his console monitoring a warp systems check for update compatibilities and testing theories on harmonics and siphon buffers. It was an exciting time right now because ship board modifications were underway in all departments, and with the energy sources in space when you need it, things were easier to get done or create because science fiction was now science fact, and most things weren't impossible anymore, just hard to learn if you had no grasp of how it works. Rachel also found that all the light balls in our sky were not all stars but actually very bright star systems with M or Y or other-class planets that read empty of any life forms or intelligent anomalies. But scans concluded that some of them had a more accurate reading that wasn't feasible at the moment, but they were very few in number. Just then Pierre spoke up about warp theory and opening worm holes across the galaxy, and he threw in a few theories on it, when crewman Wang jokingly called out something about opening up to a lower dimension where people live in chaotic blood orgies galaxy-wide, and then died and were reincarnated over and over to keep the cycle of psychosis in check. This was based on a horror movie he'd seen, which made a few ladies cringe. Mr. Gagnon agreed that there were lower dimensions, but getting to them was a whole other story. Not that any sane person would want to go to a lower dimension, but the thought was disturbing enough as a possibility. Who would want that really? He scowled at the thought, which made a few crewmen laugh.

Rachel, now intrigued by his squeamishness, asked with a gleam of mischief in her eyes if there could be a dimension in which all systems and planets had people who participated in hot orgasmic orgies all day and worshipped pagan gods, and how would they get to one of those. The bridge chuckled as crewman Gagnon's face went cringe at the thought. He cried out in protest that such a thought was provoking enough. Yes, it could happen, but that possibility was best left alone; he was still obviously disgusted by the notion. Then she called him on his un-French attitude toward passionate carnal affairs. Everyone burst out laughing and said that he was the only uppity tight Frenchmen they'd ever met. He only went red in face after everyone questioned his "Frenchi-ness." They said he might be an alien posing as a Frenchman spying on the *Valkyrie* crew, as if he were really French, he'd be going on in lectures of passion in the old country.

"Heyyyy," he argued, "that's a French stereotype done by Hollywood. I'm no prude. I keep my passion and lust to myself and those I have relationships with. You kids watch way too much TV."

"Ohhh, definitely a prude," called crewman Wiles.

"No, no, I am not a prude, I have respect for the passion to not indulge every aspect of it all the time or talk about it." He defended himself while blushing.

The ladies were now laughing as they were getting him deeper into the joke. "Yeah, usually Frenchmen are loud and open with their sexual passions, to show that it is natural and not to be feared. Hey, that's cool and progressive and all, but at least be honest about it," called out the pilot. She then mentioned that Pierre had to definitely be a clone disguised as a Frenchmen, or an alien spy, maybe for those humans on those military vessels, or maybe from another dimension where people are always so serious and uptight. The other security officer was laughing at his post, just taking in what was being said and seeing Pierre having to deal with the female crew members poking at his reserved lifestyle. He was a nice guy; he just took everything so seriously.

Just then a navigational transponder alarm went off: *BRRRRRRRRP! BARRRRRRP!* Rachel shut it off and looked at the navigation sensors. There was indeed something half a light year away closing in on their coordinates at a slow warp speed; it had a transponder like the vessel in Earth's orbit but in a code sequencing she'd never seen before. She blinked a few times, now frozen in her seat, and her heart was beating a little faster. She ran a few more refreshed scans on a fresh channel and got the same results as before. There was a ship out there, and they were headed this way. Their readings slowly came up as Roman numeral digit, like letters and numbering, followed by an energy source and wake that *Valkyrie* and the mystery earth vessel leave behind them, or like a boat or airplane would do. It was definitely of another source. It was a ship with life signatures emanating from the systems sensors. It wasn't acting hostile, and there were no energy charges or build-ups. Just a slow-moving ship with high heat signatures. She took a wild guess that it was a ship in distress. She was hesitant to move let alone speak, and she had felt a comfort just staring at the screen and running scans on it over and over, thinking it could be an error. But she knew this was no glitch and needed to bring it into attention. She slowly snapped herself out of it and called over Pierre to take a look at the readings. He was also taken back as the pilot explained that she had refreshed the sensors before each rescan at least five times, and the same readings kept popping up.

This was a shock to his system, and he ordered, "Wake the captain; this could be a first contact situation if that ship keeps its course." The security officer got up and ran down the hall to wake the captain and commander and tell them of the news. She sat at her console watching the ship get closer and closer while in disbelief. She could see that it was in trouble and could probably sed *Valkyrie* and the occupants. She had a feeling that it would probably be asking for aid; it wasn't a negative feeling. She got gut feelings at times that she never ignored, as they were usually right. Whoever it was wasn't changing course or speed but staying and then sending out a message in some foreign text that she'd never seen and truthfully had no idea what it meant. It showed up as a file in the ship's messaging video/text inbox system. She looked back, perturbed at the thought of how they'd just sent a live feed message that she couldn't decode and bring up onto the console to play while the bridge crew looked up and listened in awe with eyes wide like dishes and jaws dropped. She was frustrated, as she had no idea if she should come to a stop or keep going. Pierre wasn't really a captain, nor did he have the experience, but he was in charge until the captain or commander relieved him of that duty, and he just remained silent while running a few scans on backup systems.

Just then, Kate Braxton walked onto the bridge, half asleep and groggy, but she looked pretty awake when told the exciting news. "Ms. Blackwood," she ordered, "situation report please, and quick."

"Right, well, all was quiet until this reading showed up on my nav sensors, and it shows this object is another ship. Look." The pilot pointed out the readings on her screen excitedly as Kate was now waking up more by the second. The news must have taken her by surprise, because now she was more alive than ever and started giving orders, which began with, "Come to a full stop and raise shield harmonics." Then she got onto the ship's intercom system and disturbed the quietness with a code yellow and a possible first contact opportunity. The commander entered the bridge, excitedly asking what was going on. The pilot told him of the alien vessel on a cross course with them and explained her readings. They agreed that it had to be another ship like *Valkyrie*, not much bigger and not much smaller. Ahote was now frozen, eyes as big as dinner plates, but when the pilot showed the captain and commander the alien vessel's transponder code, they were even more convinced that they were only two minutes from a possible first contact scenario.

The bridge was now silent as the crew focused nervously with heart-pounding excitement, not knowing how to feel or what to think at this moment. All eyes were on the windshield waiting for this ship to appear, as they were about to get the first glimpse of an alien vessel from somewhere in deep space perhaps. Just then the pilot console Epad beeped as the captain pushed the button and saw Torres half asleep in her quarters. "Captain, is anything wrong? Why are we stopped all of a sudden?" she asked, shaking her head and yawning.

Kate smiled nervously and requested that the engineer get to the bridge in the next minute or so, because they were about to make contact with an unidentified ship that seemed to be in trouble. Torres blinked and asked if that was such a smart idea, but then the commander stepped in and mentioned that one should always question such an idea, but since it was the crew of *Valkyrie* that would cross paths with a distressed ship, it might be in good measure to offer assistance as a first impression type thing. She chuckled and admitted that he was probably right and she would be excited and wide-eyed too.

The feed was cut as the ship's proximity sensor alarm went off and the makings of a ship were now seen in the vast blackness coming toward them. The vessel was a little larger then *Valkyrie* and had three yellow lights burning bright in a line next to each other in an eerie manner. As she got closer, the ship dropped its speed and came to a stop a few hundred feet away from *Valkyrie*.

The vessel was a light brownish colour with small octagon-shaped windows in certain areas on its hull. It had outer hull lighting that seemed to make the ship more visible, thus exposing certain details on the skin of the hull and her decals. From its angle it was hard to explain the shape, but the best crewman Blackwood could come up with was that it looked like the front of an almond with three ovalish-shaped yellow lights that gave off a weird, bright glow that kind of hypnotized you with odd vibes and nostalgic visions, like when you blank out while staring at something and then you drift off somewhere and it's kind of dull. That's what she got when she looked at the ship.

Elisa Torres, Chief Engineer.

The engineer was now leaving her quarters in urgency, as she really was curious and yet cautious after the ship's first contact. This time could be different if the alien vessel was friendly. She had a different spin on alien life in the galaxy: if they were advanced enough to travel the galaxy, they would be good and bad. But look at the human race and find out how many people on the planet would do the right things to survive and how many would follow through with negative actions and beliefs. It would be hard to find a number that accurately exposed the number of people with good intentions and bad in all situations with agendas aside, so out there it was hard to calculate the chances of running into someone threatening. Aside from that, it's smart to be ready and expect the worse, because out there it was a whole new ball game. There were risks and dangers to be exposed, and people you don't want to meet or come into contact with, but then again, think about it: it's necessary to human evolution to expand beyond borders and reach out and explore a whole new existence outside of ours. This was another realm to human existence, and it's necessary to come face to face with hostile races in order to learn about threats and survive and grow in space and on Earth. So regardless, it's better to be prepared and be mindful that good contact could come about anytime too.

She kept walking and seeing other crew members happy and hard at their tasks. It was a weird vibe in the ship that made everyone upbeat and in better moods all the time while keeping crews calm, knowing that they were at a possible first contact. She heard crewmen talking theories about who and what they'd be meeting and what might happen next. Elisa laughed to herself as she walked by, smiling and greeting others and hearing ideas. Some thought it might be other humans come to take them back to Earth, or worse, destroy *Valkyrie* and kill every crew member in the vacuum of space. Her eyes furrowed in disgust at the mere thought of *Valkyrie*'s destruction, but she kept walking until she entered the lift and ascended the two decks as it came to a clicking stop. She walked quickly down the corridor and onto the bridge seconds later to find the bridge crew and security gathered around the pilot's console and staring at this weird looking ship with a weird and unnerving glow coming from its yellow nacelle lighting. She also noticed that the ship was venting some sort of yellow cloud into space from a few areas of its hull, probably a gas of some kind, or maybe plasma or oxygen. The clouds exuding from the vessel were thick and very dense; it looked to her that this ship was in critical condition and needed help.

She stopped short as a weird hologram of a symbol appeared in the middle of the bridge, where the Earth space vessel captain had stood when they left Earth. Looking out of the windshield, she saw only deep space and a black void. "Commander, I'm here as requested," she said. "How can I help?" She slowly walked up to the hologram symbol of a red eagle or hawk of some kind, probably a bird alien to Earth. It was holding under it a star system of some kind, like you'd see on a star map or solar system model. You could tell that the bird had to be some authority over a whole star system, which meant that this could get very interesting and maybe tense in the minutes to come.

Suddenly, Ahote shouted out, "Elisa, stay back! Everyone move from the consoles back to the banister." Torres was now mesmerized by the hum from the yellow light and the symbol. She was almost in a trance of some sort that was overcoming her slowly but more rapidly as she allowed it to.

Just then a disembodied voice was heard over the ship's speakers. "Attention. Attention, human vessel. We are under a white banner and require assistance on mutual terms as to the Federation humanitarian code of grid sector 009.86 colonies." It was loud and clear for all to hear. Then the voice continued. "My name is Captain Centaur Romeius of the Romeian Empire, and I am requesting a ship-to-ship beam aboard for a short meeting with the captain of your vessel. We will wait for your response." The voice was soft and masculine and spoke with authority and confidence, but also there was a hint of respect and panic in the tone. Kate Braxton turned to everyone else and shook her head, asking the crew what they thought. Should they beam onto a strange ship with a race so technologically advanced? Or beam a technologically advanced stranger form another world aboard their ship? Everyone looked to the security team for recommendations, but the leader, Dan Morrison spoke out.

"Well then, it depends. How do you want to play this out, Captain?"

"Peacefully yet cautiously and ready," replied Kate.

"Ok, we could board their ship and risk getting taken or something. We should remember that we're not armed and this is a first contact situation, right?" He stopped to breathe and then went on. "Or we can arm our crew with whatever they can muster up. We'll look under-armed but ready to act in a moment's notice, and we don't look warmongering or menacing."

The crew ship-wide took three minutes to all blurt out options on whether or not to let the other vessel beam down a few commanding officers first and then talk in the briefing room. If all goes well, a small number of crew in the mess hall could make it a formal gathering. It was settled after four minutes, and the crews voted on the EPads to beam aboard the captain and talk business.

The bridge crew knew the situation but also that the ship and crew were under distress and needed help. First contact was important for them and every member of the human family back home, but they also knew that if a small number came aboard, they would have a chance to fight off an invasion maybe.

Torres was coming out of the trance-induced blackout and suddenly digested what she was hearing—talk of beaming aboard strange people from other worlds and making first contract with the crew. But then she thought armed security was a great idea, as long as they equalled the alien party, and guests and crew had quick access to some sort of weapons in case they were needed.

The captain was smart to ask the bridge crew their opinions before making a decision. She nodded her head in agreement and spoke out in support of the second option. "Well then, I say let's do it. You can beam over with a small entourage and we will talk, Centurion Romeius."

Three figures materialized on the bridge immediately and greeted everyone with a Roman Empire salute, fist on their chest and the arm stretching out in front of torso, which made the bridge crew freeze in their seats and at the same time chuckle in their heads at the salute's dramatic sense of identity. Yet they kept respectful at all times. There were two males and one woman. The males were tall and slender but fairly muscled for thin warriors, and their necks were longer by only a foot than a normal human. They were 6' tall with pale skin and jet black eyes that made the crew suddenly apprehensive and perturbed deep inside. Their hair was shoulder length and straight in a ponytail. The woman was about 5' 9" with an athletic figure and the same jet black eyes and elongated neck as the men. Their ears were just open cavities covered by a small flap of skin, and this female had a pale tone of olive skin. She also wore what appeared to be cosmetics over her eyes and red lipstick. Overall, their appearance was disturbing at first glance, but their demeanour and polite manners were more apparent, which put the crew at ease. When they smiled at the captain, their facial features were a little tight, one could say, which made them look a little devious. But Kate knew that judging people at first glance was not a logical thing to do.

"Welcome aboard the civilian spaceship *Valkyrie*," uttered Kate with light-hearted emotion while she shed a tear of happiness and joy. Despite the looks of these people, she felt no fear from them. For some reason, she was going to go with it, aware of any moment or movement. Kate was feeling joy as if she was meeting a long-lost loved friend again. It was a home-coming feeling deep in her stomach, and her head felt light and fuzzy energy. She wanted to cry, but she bravely held it in and put on a smile. She had to fight to hold back her tears of relief and joy for now and be a captain on a first contact.

"Well, hello, Captain Kate Braxton and crew. My name is Centurion Romeius, and this is my ambassadorial officer, Nomen, and my first officer, Luandria. Do not be alarmed by our appearance, as we are highly evolved and our world is in a dimly-lit star system. Our eyes have adjusted to our environment; we come in peace and have no quarrel with you." Romeius spoke enthusiastically and held out his hand to the captain with a disturbing yet friendly and polite gesture of openness and trust. Kate took his hand and noticed his skin was a little rough, like a male human's hand. At that moment, the bridge crew let their guard down a little bit and smiled with the Romeian crew, who were happy to meet them it seemed.

Ahote inched up to the captain and introduced himself as *Valkyrie*'s commander and a direct descendant of the Pledan star people, the Indigenous folks on Ancient Earth. The captain held out his hand and squeezed kindly, with a smile on his face, replying that it was also a pleasure to meet them and he was impressed to see a Pledan on Earth with other Pledan descendants of the North and East worlds. He commented that these kids must be proud of themselves to be out there, but unfortunately, they weren't there to engage in any relations campaigns at this time. He said it was an honour to meet peaceful Earth humans and not human military factions. They needed help because their ship had suffered damaged in a battle some twenty thousand light years away against the Earth human military ships. Something went wrong with the weapon energy transfers and threw their ship across the galaxy almost, plus their beam drive was partially knocked offline and food supplies were at a dangerous low. They would only last two weeks at the most if they couldn't find an M-class planet by then. They needed assistance before life support failure in two months.

Kate was now intrigued at her new guests and had the gut feeling to ease up, because no one at that moment felt any reason not to trust them. She asked them to join her in the mess hall for a formal meal and drink while they talked about what was needed and how soon they could start repairing the other ship. The Romeians appreciated the offer, and the woman responded with a polite smile and said, "Thank you very much, as you Earth humans would say, I believe?"

Kate smiled and replied. "Yes, we do. You know about us then?"

The crews started to walk toward the door to exit the bridge when Kate ordered Kyle and another member to man the bridge and excused the rest of the crew to follow. They walked out the doorway and into the hallway while Ahote told them the story of how this ship came to be and how he came to know these brave individuals and join them in building the ship. The Ambassador cut in and asked, "You are what they call … Native American, yes?"

The commander replied with a smile. "Yes, I am."

"Ahhh yes, your people have had a rough past with your governments, but you have managed to survive to this very day. That is warrior of your people; this I commend." Ahote chuckled and thanked the Romeian for his comment as they entered the lift, but it wasn't big enough for all of them, so the captain and guests went up first while the others caught the next one. Kate explained how the ship was run, pure simple and primitive to them, with a laugh, which made the Romeians laugh also. But when they came out into the observation lounge, they were amazed at the room and the design and the view.

"Wow! Amazing. Not only engineers but designers too. Captain, I must say right now that you and your crew are indeed incredible and intelligent Earth humans. I must hear more of how you have come to be, in full detail, though." The Romeian captain's raw passion and excitement were kind of weird to Kate, but she excepted it, as it could just be that they weren't used to seeing young adults, and the crew were all university students, none older than fifty, and civilians with no military training. It should be seen as an honour coming from members of a galactic race.

When they walked into the mess hall, they weren't prepared for the massive, curious crowd waiting for them. All the ship's personnel had snuck out and were standing around, waiting eagerly for this moment. The crews' faces weren't indifferent when the Romeian crew walked into the room with smiles and nods and a few hushed whispers here and there. Overall, there were teary-faced smiles all around while a few women started to weep silently. The mood in the room was one of deep feelings of homecoming.

Cassidy was preparing a table with some hot coffee and a few platters of assorted foods and pastries. "It's ok, Captain, I saw the ship approaching and had a feeling. Welcome to our visitors, and welcome to Cassidy's Intergalactic café." She nodded her head and held out her arms with a face that looked like she was going got bawl her heart out. At this point, how could you not cry, meeting people from other worlds who were created like us and vice versa? They too started off as infants.

The Romeians' faces were happy at this welcome; their smiles grew big, and that was enough for poor Cassidy to be mortified inside while still smiling and emotional. She didn't mean to be put off, but it was just the way they looked. She laid some cushions on the seats and helped them to the table and served them a little chicken noodle soup. The Romeians gratefully accepted their meals with curiosity and studied it when their captain picked up his spoon and dug in and tasted the broth. They gleefully agreed that it was tasty. Then he tried some chicken and began to gobble it down as if he was really hungry, but in truth, he liked it. The others took to eating the soup and also enjoyed it. Then Romeius spoke up. "Yes, we have heard of your race, Captain, some good and some pretty crazy things for sure, but I can't help but feel that you're all not as ... psychotic as we've heard."

Kate spoke up and gave them a brief but descriptive history of humans, and then highlighted where humans were now and the waking up that was going on and which had led to this point with them talking over a bowl of soup light years from Earth.

Then the centurion said something that caught Kate's attention. "Then it was and is your military and government who have a space fleet that goes around doing the Dracons' bid—." He stopped and paused, biting his lip as if he'd slipped out something he wasn't supposed to tell them.

Laundria, the first officer, chimed in. "Tell them, Romeius. They made it this far, and whatever they're looking for, they'll need some knowledge. Obviously they're just civilians who found a way to make space travel possible for themselves, plus they're new out here and know nothing of what's ahead of them."

Romeius smiled and looked up and continued. "Like I said, your military has a space fleet that flies around patrolling galaxies and star systems for a race called the Dracons, who do not like humans, unless they fall into line with their agenda on Earth and afar." He breathed then continued. "You humans are troublemakers out here, and you're wanted by them, you know. But not to worry, as the Dracons learned a long time ago that the Romeian Empire does not answer to them; let's leave it at that and the fact that neither do we listen to your military's fleets, so we shall not turn you into them. But beware, as others out here just may if they are desperate for food or money, or just in league with the Dracon Empire."

Word was spreading through the Earth Space Fleet and Dracon sub space communications that a human civilian vessel had escaped Earth, and the crew was wanted alive immediately. The mood around the room was now quiet and tense at the news that their actions had stirred up such a commotion. The authorities back home had told families that their kids were dead. Such lies were quick to anger some of the crew, while Kate kept a stern composure.

"Captain, do you and your crew have a destination or mission, or are you just exploring?" asked the Romeian first officer in a kind tone.

Ahote chimed in. "We're on our way to the Pleiadean system, actually, to check out an anomaly we saw two years ago. We're curious to find out if the Pleiadean star people really do exist and make contact hopefully."

The first officer breathed in relief. "Ohhh good, they are with mission and not totally helpless—I mean aimless with no real goal. Excuse me, I don't mean to sound rude. It's just quite rare to see young earth humans accomplish all this with your government's control over you and no help from them." With that she smiled and told them that the Pledan system and its people were indeed real and of all skin colours, much like the crew, but they were united by a Galactic Federation that looks out for each other, monitors primitive planets, and keeps the peace. They already knew of the *Valkyrie* and her crew.

The Romeian Ambassador spoke up. "I think it would be fair to tell you, since you have achieved all this and made it this far, that the Galactic Federation is not what it all seems to be. They are corrupt and don't allow Earth humans on any councils, nor do they allow human ambassadors. They do not have your best interest at heart, so take heed with them."

Ahote wore an astonished expression and was frozen in his seat at this news. The female Romeian officer smiled and replied that they weren't all corrupt. Kate told them that they had planned on requesting asylum or protection on the Pledans' home world anyway. The crews gathered excitedly around the table in disbelief, listening to the story about a not so peaceful Confederation of space-fearing people who have existed for over one million years. Then they learned that although the Romeians don't share membership in the federation, they have no ill will against the federation, or any of her star system members, either human or cat. Both parties agreed that peace between the Romeians and them was important. It was stated that only the human races were mainly on humanity's side, and a few other animalistic races were dropping out of the federation and siding with the Pleiadeans to sympathize with humans from all over the galaxy. But few humans from out there did join factions of pirates, smugglers, and rebels—your bad minority among the humans.

Kate asked the guests about the Dracons, and the Romeian captain explained that they were a very ancient race of reptilian royalty and warrior beings who were hellbent on enslaving lesser races for their colonial expansion. Earth was always a target of interest in her ancient days, back when the Romeians were under her rule as the Roman Empire on Earth. Then one day the Roman Empire was told to leave and another ruler was sent to earth to take over. That's when they installed the Abrahamic religions over all pagan beliefs. Many ancient Earth libraries were destroyed by the new King of earth and his new religion to enslave the masses. Now the Romeians don't serve the Dracons or their crony criminal imperialists. Believe it or not, they hope the human race finds its way from Dracon control, even though it's highly unlikely based on humanity's own efforts.

The room was quiet, and so was Cassidy, who was somewhat horrified to hear this information about Earth and part of its history, but then she thought of Dick and Marie and kind of chuckled when she pictured them as reptilian alien overlords. *Suits them*, she thought. Then Romeius asked Kate the story of her and her crew and how they got there. It was a great story, as the *Valkyrie* crew took turns explaining the event with the new astronomy students and the planetarium telescope and the Pleiadean lights, and then the plans Kate drew up and the school revolt in class. They told up to the point of them launching the ship and the encounter with the human spaceship in orbit over Earth, and how they emp charged an Earth-like military vessel larger than their ship and got away. They told of the compelling feelings inside of wanting to join the stars and explore and meet new peoples they believed to exist, and that their hearts had hurt so long just to be in this ship and reach this day. They described building the craft and all the non-believers and protests, and then the dedicated students who wanted to see if these Pledan star people actually existed. This had motivated the best of the minds to actually unite and do what seemed like the impossible. They also wanted to see if their space travel theories were true.

This made the visitors very excited indeed as they cheered and ahhed and oohhed at the crew's resolve to leave Earth, and all the adversity they'd crossed. This also made the Romeians more compassionate to their struggles and victory. "Well, well, Captain and crew, you are an amazing group of people. Most of your kind are still held in bondage it seems, but you managed to defy your masters and do what you set out to do. This makes you warriors in our books, especially because you will now dislike your government's military as much as we do."

Then the ship's security officer spoke up. "Believe me, we don't like them already. They were the ones calling the Pleiadean lights a military experiment and took all the credit for it." A few other crew members piped up in agreement, leaving the Romeians to laugh and sympathize with the crew at the same time. Cassidy came out with a platter of veggies, which the Romeian visitors liked. She asked them about their culture while she took a seat close by and listened intently as they explained that they were also an empire that just kept to themselves. They'd learned that life as rulers under another empire's orders wasn't a good life, and they found that they hadn't wanted to rule over others after years and years of Earth conquest and fighting the natives of all lands and their planetary spiritual gods. So they came back home to Romeius and planned to leave the Dracon Empire but had to fight to get free, which cost their people many lives for their freedom. Then a weapon was created and the reptilian scourge backed off and signed a peace treaty to never tangle with the Romeian people again, in exchange for a promise that the weapon would not to be released. Not even the Romeians wanted to launch the weapon.

They don't run on a money system but a point system. For days they work to provide for their empire, and all are taken care of with no poor or homeless, very little crime, and a prosperous government and people. Their main crop was veggies and fruits, and they didn't eat meats often, only if they needed a certain protein in their body not know to Earth humans. They grow up to live at least five hundred years.

Kate was asked if they had means to protect themselves. She responded reluctantly, not wanting to give out too much information, but reported that they weren't defenceless but not in a position to be a warship either. Then the Romeian captain offered them captured rifles, pistols, laser beam weapons, consoles, and recipes for supplies and fresh Romeian cuisine in exchange for an engineering team to help them repair their ship and life support. He explained that he had an interest in the ship and crew and wanted to keep in touch as mutual friends to exchange news and stories of adventures. He might also have some work for the *Valkyrie* crew after they got settled on Pleiades. The crew gasped in unison, and the captain and commander froze in their seats in disbelief but also relief over an idea that was currently on the ship's engineer's console.

Torres stepped up and said, "Captain, I'd like the honour to be the lead on this project. I feel I can learn more about spaceship engines, and that will benefit us later on and maybe even improve *Valkyrie*."

Kate smiled and looked at Ahote, who also chuckled and nodded. Captain Braxton spoke up. "Well, I think we have deal, Captain Romeius." From there the night took off with a party and celebration like no one on the *Valkyrie* had ever thrown.

CHAPTER FIVE

Two days later, 08:00 hours, 102 days away from Earth

Torres was in the engineering bay on the Romeian vessel working alongside a Romeian engineer running diagnostics on the now fixed transwarp drive core, which ran a type of quartz from the Romeian home world that was naturally a blue and transparent quartz rock. If you looked at it for a while, you could get lost in a trance state and come to an hour later feeling rejuvenated. It was also used as a stress reliever for all Romeian citizens who were healthy and abled in their society. But when the quartz was utilized in the transwarp conduits, the colour was a hypnotizing yellowish light that she had seen when this ship showed up two days ago.

Elisa and the female Romeian crewman Loyian were laughing and comparing engineering bloopers or near misses. Even though Elisa had had a few minor bloopers that could have delayed *Valkyrie*'s trip through the cosmos a little, they were never in any danger. She learned while helping the Romeian repair their ships that not only did they use their quartz as fuel, but the ship's hull started out as an asteroid. Then the rock was moulded into a hull shape and the interior was drilled out by carving drones that mined their way into the rock, thus making the asteroid into a space vessel. From there the Romeian people finished the interior and the technologies.

Along with learning about Romeian technology, which was at least five thousand years ahead of hers and consisted of more advanced and different metals, she learned that Romeian women enjoyed the same rights as men. They were a pagan people, and both men and women could have up to twelve different husbands or wives, and vice versa if they chose. They were warriors when times called for battle, but they mostly kept to themselves and took care of their own race, which included all coloured Romeians from around their planet. No matter their geographical region, all were Romeian and enjoyed the same rights and duties. They had no poverty and a low crime rate, even though they were an empire free to be Romeian with Romeian ways.

The Romeian people used a credit system to make life affordable, and they capped costs on everything so that no one can dupe or exploit others. It worked, with an over-abundance of resources to go around. They were an offshoot of the Roman Empire but now more planetary nationalistic and into military service and patrolling other star systems with territories and colonies held, which for security reasons were never scattered more than four hundred light years apart from the Romeian home world. These warrior people were not meat eaters but big on drinking what they called Romeian wine. It was super potent alcohol that was ten times better than the strongest human-made liquor; too many cups of wine could kill a human being due to alcohol poisoning.

The two ladies laughed and chatted as Torres told the Romeian crewman about Earth and their customs, religions, and timeline of our insanity to now, as we are still in some sort of bondage politically. The other girl listened as the two stood around the console talking about how the crew had escaped and become a little more evolved than they were years ago. Until the event at school, nobody knew that this talented crew existed, or that they'd make it this far with *Valkyrie*. She explained that human will and belief goes a long way when humans actually set their minds to large tasks and create the impossible. "I mean, you can ask me two years ago from today about building this spaceship and overcoming my fears and insecurities, my hopes and dreams, and then ask me about her now! I would have never mentioned this or even known what I was capable of. But then Kate came around with her ship plans, and I had to rise to this challenge."

The Romeian smiled and nodded her head proudly at Elisa's comment and agreed. "Yes, you should be very proud of yourselves. Excuse my saying, but last we checked on the human race was when your countries went bankrupt, as you call it, in ... ohhh, 2008 you called it?" she breathed and then continued. "Yes, we left you a mess since the Roman Empire, and we could still see the remnants kept by your leaders. I'm so sorry for our direct influence on you Earth humans; after we left Earth and left you to the Dracons, we always had a biting feeling in the back of our minds that we did help in ... umm, sabotaging your race when we ourselves wanted to be as free as you on Earth now fight for."

She looked down at the floor and nodded. But before Torres could say anything in return, the two engineers were interrupted by the systems check alarm alerting that the scans were completed and the warp core and systems were all set and ready for departure after a short system reboot. The girls both smiled at a job well done, and Torres replied, "Don't worry, this makes up for it. Kind of." The two laughed unison while the other Romeian engineers watched with smiles on their faces at the human and Romeian crew getting along quite well.

Captain's quarters, 14:15 hours

"Well, I'm glad we could share information and art with each other for more needed information on this galaxy, and all this about the races and factions ... wow!" Kate blurted out as Commander Ahote and Romeius smiled kindly at the intelligence and data that was just downloaded into *Valkyrie*'s computer systems, including the Epads. They were given information on all the races the Romeians had ever come across and knew of. Some information was held back for security reasons by the Romeian trade and intelligence council for planetary and trade security, which Kate understood fully. She assured the captain that it was all right and they didn't want to overstep their bounds, so everything given to the *Valkyrie* crew was much appreciated. She had skimmed through horrid pictures of certain hostile races and beautiful looking humans from not too far away from *Valkyrie's* current location.

"Well, Captain Braxton, I see you now have a full library of knowledge to get you through your trip to Pleiadeans. Is there anything more you require?" Ahote smiled and asked

"Well, what do you say about ship board security out here?"

The Romeian captain smiled and laughed out loud in a theatrical way, as if the commander had asked the right question or hit some idea on the head.

"Ahhh, you are smart indeed, and careful, as you should be. Listen, Captain Braxton, we are a people who run on merit, effort, and loyalty, among other things, but you and your crew have displayed hospitality and control. You've showed us kindness, not fear or distrust. You gave my crew good food and lent us facilities to use when you didn't know us that well. Consider this all a gift, friends to friends, hopefully. And with that, I will throw in a few extra dozen Romeian phase plasma assault rifles and pistols for your crew. But I would recommend upgrading your hull armour and shielding so it's impossible to just beam aboard."

Kate's eyes opened wide like dinner plates when Romeius explained that it was not every day civilians from Earth built a starship to cruise the stars on a mission worthy of such an endeavour, and this was an occasion that deserved merit and a gift for this fine crew. He then acknowledged that some of her crew did a good job at not being horrified at the appearances of the Romeians and their smiles. Ahote reminded him that they were of a different mind, but now that they were up in no man's land, this was a game changer that had taught them that no matter, this was serious: change our ways or be lost in space with no way of getting home. It was a risky mindset also for the other races out there that meant harm to others.

Romeius explained that not every Dracon race they'd come across would be hostile to Earth humans, and not to expect all reptile races to attack *Valkyrie* when she came across them. Then he smiled at the commander and stated that his ancestors were also out there and knew about *Valkyrie*, and that Kate and crew were coming.

This was a moment breaker as Kate and Ahote froze at Romeius's words. "Wait. The Pleiadeans? They know we're coming?"

The Romeian captain sat back in his chair and spoke softly. "Yes, this morning my commander reported to me that the Galactic Federation had a scout ship orbiting the other side of Earth when the *Valkyrie* left the planet and ran into the human space vessel. The Pleiadeans tracked the *Valkyrie* heading in the direction of the Pleiades system, plus the comm chatter they were sifting through picked up on some law chatter that mentioned school students building a spacecraft to some star system." He breathed and continued. "They know you're coming now and have put out a message to all Pleiadean ships to keep an eye on *Valkyrie* and observe her crew. They are to assist if necessary and determine if the *Valkyrie* crew were not hostile."

Ahote blinked and smiled, thanking the Romeian captain for his news that seemed to impress Kate, who shed a tear. Romeius caught Kate's emotional response and gave her a kind yet trusting smile then nodded. "Well, I'd say it's about time to have our departure drinks in the mess, eh Captain? Then we must depart and get home as soon as possible. You and your crew need to safely make it to the Pleiades as soon as possible. You're almost there, so keep going."

The commanders stood up slowly, smiling in satisfaction at the conference and information swap that just went down. Kate led the men out of the briefing room and down the corridor toward the lift. A few *Valkyrie* crewman with three Romeian males walked past them, smiling and greeting them as they walked by with glee, all grinning, chuckling, and sharing stories of friends doing crazy things back home. This made Kate, Romeius, and Ahote chuckle too. The Romeian captain mentioned that he was glad that the emergency assistance turned out exemplarity for the circumstances and was joyed to find such an adventurous and daring bunch of Earth humans exploring space for the first time ever. He even commented on the crew's engineering and interior design, and how humans had a huge capacity for imagination and creativity. Most humans he knew about on Earth were still enslaved by the old empire mindset that still ruled over the Solar System.

The conversation seemed to take a heavy twist as the Romeian captain briefly explained again their part in enslaving the human race on behalf of the Dracon Empire, and how they have checked on Earth from time to time to see that humans were still under the rule of politics, money, and religion. Then he reminded Kate and Ahote to be mindful from now on about who they came in contact with, reminding them of the data he gave them and suggesting that the crew study the libraries as soon as possible and watch the instructions for rifle use.

Romeius was a noble and fair man, yet he had his ways and orders, his likes and dislikes. He told it as it was, and Kate came to like him for it. She also found the Romeian visitors' company provided a bit of security, but he was also frank at times, like now, getting serious again hours before his departure. She could see that Ahote had picked up on it, but he didn't show any signs of concern or worry on his face. That was a reassuring thing.

They boarded the elevator as Romeius continued to say that the crew should expect the weird and impossible to the most ironic and familiar things to happen, as life out there was very rich, with some sectors of space littered with planetary colonies and others sparse with colonization. Some species would be terrifying and others would be calming, but most preferred to stick to themselves or their factions away from the Dracon and human space fleets. But they weren't without thought for other peaceful peoples in need of assistance. Kate smiled and nodded. Her heart warmed inside again, and she swore that right now she could cry happy tears and relieved emotions, but she bit her tongue and thought about the quick glimpse of a Dracon she saw earlier on her Epad, which seemed to wipe the tears and sappiness from her face.

Cassidy's Intergalactic Café

Cassidy watched the crews of both ships walk into the mess in groups, laughing and talking excitedly amongst themselves, but she noticed it was the *Valkyrie* crew doing most of the talking, with lots to say and ask. It was funny that over three days of these Romeians being onboard, the crew never ran out of questions or stories to share. It was as if they'd held on to all this for so long, and now they had the chance to blurt out everything they ever thought about the universe and life in general. She turned around to see the Romeian chef from the other ship finishing cooking his meal he'd prepared for this drink-fest. It was Romeian burgundy falk peppers on Romeian rice grain that looked like white rice. It was the same size, except the whiteness was natural and healthy to the Romeian diet. The peppers were a combination of an eggplant with green pepper texture and size in a pear shape. It tasted sweet, she admitted to herself as she swallowed the aftertaste and noticed that it tasted like an eggplant fresh from the ground. It was healthy for most Milky Way races and had a cousin vegetable on other planets throughout the galaxy. Cassidy liked the taste and texture too but wasn't a fan of the rice grain, as she found it too mushy. If she ate rice it had to be cooked, not too dry but not mushy either. It was ok because she had lots of normal rice back in the food storage if need be.

The female Romeian cook had been recommended by her crew for the evening departure event to be thrown. The feast and drink had been planned and finalized by the Romeian first officer and Kate, and like always, Cassidy had also been drafted to prepare some coffee and a dish of her choosing that represented Earth's healthy but yummy food. This made her proud, which drove her to skim the cookbooks she'd brought until she found it. It was perfect now, because Cassidy made a chicken Caesar salad with some homemade garlic bread and wine. The two cooks sampled each other's food with great joy and excitement, and they were oblivious to the crews around them making drink requests.

Ten minutes later the last of the crews entered the mess hall and gathered around the tables, which were specially shaped together like a summit conference round table that surrounded the café counter. Everyone's eyes turned to Cassidy, while Kate and the Romeian captain were the last to walk in. They made their way to the middle of the tables in front of the café counter top, and Kate spoke up to make an announcement. "Attention, everyone, attention. Good. Welcome to the final day of our first contact, as our new friends will be departing today to go back to their home. We did a job well done, and because of that, we've been invited to this feast as friends of the Romeian ship's crew. But one thing is for sure: we must be doing something right."

Romeius stepped in, clearing his throat a few times, "Yes, we again thank you for your trust and friendliness in our time of need. You opened your ship to us and didn't even know us, but you took that chance, which is different from most humans we find from your Earth space fleets. We are greatly appreciative, and as your captain says, friends indeed to the wonderful crew of the *Valkyrie.* May your voyage be everything you expected, and be safe." With that. he raised his empty glass and then grabbed the bottle and poured some Romeian wine into it. He took a sip then passed the bottle around for everyone to pour and join in the toast in unison as the crews all held up their glasses while the bottle made its way back to Romeius. He then poured another glass and raised his with everyone else in perfect timing. "To new friends from afar and possibly peaceful relations." Everyone saluted and drank the dark red alcoholic beverage.

Kate found the wine actually quite sweet and not too bitter, as did Ahote, whose face wore a blissful expression, as he obviously enjoyed its sweetness. It was like a grape juice of some kind, and he drank his glass quicker than Kate did. He smiled a satisfied and happy grin toward the Romeian captain, who had drunk his in two gulps. This made some of the *Valkyrie* crew scream out and cheer like a bunch of frat party people do when someone does a shot of some kind. Then came the laughs and cheers to the Romeian captain, who looked at the captain first with a confused gaze but then looked back at the partygoers, slowly getting the hint of cheering and drinking before he held up his glass and shouted, "ROMOKA HAAA!" He cheered himself, which drove a frenzy of crews hugging in groups, patting each other on the back, and pouring more drink and passing it around sparingly.

Some shared cups in brother and sisterhood, and for the moment, Cassidy lost herself in a bizarre moment that made her so happy inside yet amazed and in awe. She realized that this crew had come a long way and out of their shells since day one of the event at the university observatory two years ago, and it made her proud to see a first contact and a friendship with this star people. They were authoritarians but nationalistic on their own, not harming anyone else. But then again, they did seem like a people in arms defending themselves from a bigger threat everyone was starting to talk about now. Then an idea hit her. What if they were getting themselves into something deeper than anyone could have thought, like space being hostile with bigger fish out there who mean peaceful people harm beyond anyone's comprehension. They needed to be very careful and prepared for this. After three days with the Romeians, Cassidy had learned that all her life they'd been lied to, and the truth of life and existence was a lot more diverse than anyone could have imagined, and now she was really comprehending the reality that space was filled with all kinds of people they were yet to encounter—some bad, most good or benevolent. They needed to be prepared to see strange and unusual things. People evolved from typical humanoid forms, and this Earth secret space fleet, as one Romeian told her, was a military multi-nation space force hellbent on military strategic control, mainly in service to some bigger force more sinister yet so advanced most humans would drop dead upon first knowledge of such a race. Inside her heart raced and her tummy went tight when she swallowed a pit of concern, but it passed a few minutes later.

Security Room

Sherry Wiles sat at her console watching the security teams help the Romeian crews disembark their tools and belongings from the three-day stay. They walked to the operations bay, where a Romeian shuttle was to enter and load the crew back on their ship. They were also taking the time to go over the new weapons list given to the security team as a gift for defence if the crew ever got into a skirmish with some hostile race. Wiles counted the open crates of rifles that looked like AR-15s, but the ammo was a crystalline cartilage of Romeian quartz that could be recharged after a lifespan of forty-eight hours of usage. The pistols were the same design but smaller and with a bigger grip. They were red and chrome coloured with Romeian symbols on the buttock, or pistol grip. It was definitely a nice touch to the guns.

"Wow," she thought out loud when she counted exactly thirty rifles and pistols sitting pretty in the crates, which Sherry was now going to inventory and stack on the rifle racks in the armoury office. It was locked by an electronic code the security and other qualified crew members would soon need to remember in an emergency or away mission somewhere. It was perfect, as a shelf was already set up in one of the smaller offices tucked into the back of the two bigger offices, and it would be a good place to hide the weapons. The space and size was just what they needed to safely store their new rifles and have easy access to them. The rack hadn't been built for them specifically but was built to hold them securely by default. Their luck made her chuckle to herself, thinking she had it made.

Grabbing the crate and dragging it over to the closet, Sherry grabbed each rifle and stockpiled them upright in each cubby until all were tucked away. Now she needed space for the pistols. but she could see nothing that could be used at this moment. She thought of just keeping them organized in their crates along the opposite wall in that same closet; yes, they'd have to bend over and pick them up, but that wasn't a problem. Five minutes later her task was complete. She closed the door behind her and locked in the security code: 45793. The lock tweeted that it was secure. Sherry walked over to the main round console that displayed a three-dimensional image or screen in mid-air, like a hologram. People could tap the keyboard on the console as the images cycled through.

She sat down and spoke to *Valkyrie.* "Ship's computer, pull up file Valkyrie-001 and display on console table." There was a beep when a symbol popped up with the Romeian Empire logo on it. It floated in front of her as she forwarded to page one of the different races in the whole galaxy. "Ship's computer, open file Valkyrie-001, hostile races in Milky Way galaxy, page one." Just then, a horrific image of a head of some reptilian beast—a humanoid for sure—appeared. In a female voice, the computer stated, "Dracon Empire. Home world: Dracon Prime, located in the handle of the Big Dipper star constellation." The heads were huge and scaley but came in different colours and classes, from royalty to warriors and servants. They were fierce looking, like a tyrannosaurus rex with a more human shaped head but a t-rex shaped snout and jaw line. They had a muscly looking face with sharp teeth and alligator eyes that were yellow to red in colour. They looked through different light spectrums. These creatures had small flaps of scaled skin over holes where their ears should be, while some had horns on their heads. The royal class of Dracons had white scales and were at least 20" tall. They were carnivores known to eat all types of humans and humanoids, especially the young ones, because young blood was pure to them. The green-skinned ones were everyday reptiles who served and worked in sciences and markets and such, and the brownish blackish ones were the warrior class and messengers between the Empire and Federation forces. Their bodies were created for combat, and they were at least 16" tall. The brown-skinned guys also had wings and horns and looked just like gargoyles you'd see on top of buildings in New York City. They were evil looking and very ferocious and technologically advanced, with no real spirituality. They stuck to fake religions to control their colonies with their warriors.

On the next page were other Dracons who were not hostile and looked like humanoid raptors. They lived near by the Dracon system but weren't bothered by the Empire, as they had a weapon that could demolecularize ever reptile across the galaxy for protection. The Galactic Federation was mentioned as having nothing against the Gorgons, but no one wanted their weapon to be used.

Next was a surprising chapter about human beings of all colours and races united as one race. They were close to Earth in what was mapped as the "Human Corridor." It was a bunch of planets that if connected dot to dot would form a corridor, just one line that matched them all together. But she read that they were at war with the Dracon imperial forces and continually calling out Dracon abuses against Earth humans to the Federation. Often ignored, the humans decided to rise up and ally with Earth's secret space force dodgers, who went AWOL because they couldn't in good conscience keep oppressing their own people as Terran Protectors against the Dracon Empire. She suddenly felt disappointed and her heart sank. How could a bunch of advanced people in a galactic federation side with a race of human-hating and enslaving parasites?

One human race stood out and defied the Dracons and the Earth human space fleets and fought them off their planet after many long years of occupation and genocide. Eventually they freed themselves; they were called Orions, from the Orion constellation, and they were a human warrior civilization with two kinds of humans: a little offshoot from Terran "Earth Man" humans, with red to blue and pink-skinned humanoids with slightly bigger eyes and smaller ears. Their skin colours were softer and glowed radiantly, hence making them gorgeous with their hairstyles and personal grooming. They were the Orion faction of the Terran protectors and were a little more warlike and oppressive toward captured Dracons and Federation members. Orion Terrans were a heathen race with some beliefs rooted in ancient traditions. They were true to their word and cared for their fellow humans, called out injustice and dishonour, and cared for nature and lived in harmony with it. They honoured family and tradition and were quite loose when it came to lust. Another faction of Orions were shorter humanoids who looked like typical grey human hybrids. They were taller with more balanced grey and human traits.

These Orion groups had left the home world and Terran protector to pursue the Dracons and any corrupt Federation fleets shooting up and destroying Empire ships, anywhere and everywhere they could find them. They were called the Maquisaars, and they had bases throughout the quadrants where allies of the Dracons had colonies or trade routes in space. Sherry giggled and thought these freedom fighters were pretty cool people but also warriors. As Sherry was a pagan at heart, she loved freedom and knowledge and adventure. The more you travelled, the more you learned, she thought, plus she loved a good fight. When she read this, Sherry knew that she just had to meet the Orions, possibly on their way back to Earth. It was definitely a thought worth bringing up to Captain Kate Braxton.

Ms. Wiles kept reading for the next hour, studying all the races and quadrants in this galaxy. She realized that out there it was kind of like being back on Earth, where you have good people, bad people, and in between people. This pretty much made her day and put into perspective the reality of life no matter where you go. Sherry started to laugh harder and harder, until she was on the floor roaring in laughter and rolling around. The other security officers walked in, wondering if she was ok. They looked concerned until they saw the hologram library. After what seemed like many minutes, she tried to speak. "You have to read this memo they gave us. They're all like us—all of them. No matter where we go, we are everywhere." The security team leader squinted his face in confusion, wondering how that made sense or was funny, but after all that had happened during this mind-blowing experience, everyone was dealing with this reality in their own way. Sometimes she could go insane for an hour or so. Gathering around the console, they all took a seat, and Sherry took them through the information from the beginning. They too were amazed at what they were going to learn very soon.

Bridge, 20:01 hours

Kate was walking away from the docking bay down the corridor, remembering the party that just occurred. Together they drank two bottles of Romeian wine, and now half the crew was tanked, while the other half was tipsy. The few who didn't drink now resumed their duties. She chuckled and didn't mind really, as everyone had a good time and got quite loose. It was a well-deserved event, and they could afford the time off. The captain was headed to the officers' lounge near the bridge in case she was needed right away. After both encounters with their guests, she expected the neighbourhood to come knocking at any time. She chuckled out loud to herself at the thought that even in this galaxy, news travels fast. But she really wanted to use the couch and get a few snoozes under the window. She was close if needed, and Cassidy had two pots going at the small coffee bar.

She turned left and walked down the main corridor, passing some crew returning to their stations. She thought she should pay a visit to the bridge to make sure her order was given to the helm's pilot, Rachael, who didn't drink but had one shot of whiskey and partied even harder in high spirits. She wasn't drinking alcohol, just coffee, and was laughing and telling stories about her time at the helm and those long, quiet and mundane hours getting readings while on one course and speeding in open space with nothing but the stars ahead of them. This made Kate smile, as even the most boring parts of this trip were memorable, like when the girls on the bridge were heckling Pierre about romance, and Kyle was talking about alien weddings and romances and what they might be like as if he actually knew. But at times Kate wondered if he was an alien from another world.

Kate reached the bridge to see the pilot just sitting down and opening up her console computer program. She turned around to see Kate walking in and asked her captain what the orders were while they watched the Romeian vessel pull away in reverse and then turn to their port and zoom away faster and faster until they disappeared in a yellow flash of light several metres off *Valkryies*'s bow.

"Pilot, set a course for Pledan system, warp 9, at pilot's discretion." With that, Captain Braxton walked toward the small corridor and finally reached the lounge doors that opened automatically and closed behind her. The dimmed nightshade lights ran on the ship's Earth clock. She sat down on the couch and slowly laid herself down on the cushions, which were soft and light grey in colour. Closing her eyes, she tried to shut off her brain and tune everything out except for the warm humming of the ship and its warm atmosphere, to the point of a light meditation. Moments passed by into what seemed like thirty minutes when she could swear the lounge door opened. She turned her head to see Cassidy with a tray of cups and a large silver coffee thermos. She set them down quietly on the bar counter while spreading out sugars and cream. This brought a smile to Kate's face, knowing that the bridge crew was going to want some of the special coffee Cassidy blended just for hangovers or too much alcohol consumption. She still wouldn't tell the students the ingredients, as it was given to her by a coffee farmer she did business with. Fair enough. It worked, and it was to protect the farmer's farm and family from exploiters and venture capitalists seeking to capitalize on it.

Kate must have dozed off after that, as she woke up and it was three in the morning. The coffee was still smelling fresh when she opened her eyes to see David Kim and Thomas sitting at a table in silence, drinking and shaking their heads. She pulled herself up slowly, squinting in the still dimly lighted room, finally balancing herself on her feet. She walked over to pour herself a cup when the two men looked up in surprise and nodded, mumbling "morning." Kate sat down at the table and replied back to them as she took a sip and warmed herself up, feeling that kick and her tummy going soft, thus giving her a jolt of energy.

"So what's been happening on the bridge while I slept the whole night … ughh?" She still felt a little groggy when Thomas reported back that nothing at all had happened and all was quiet and calm. He admitted that the crew was more confident now, knowing they had made friends with a Romeian vessel and its crew, who helped them and evened the odds for them. There was still strong crew morale, but it was mixed feelings up and down, and often drove a few crew members into fits of mental meltdowns. But nothing too serious that couldn't be cured.

David, an engineer who frequented the bridge and mainly worked in the engineering room and dispatched round the ship to different areas to check up on systems or whatever was needed, stated, "But what I can say, Captain, is we chose this adventure, and we've been talking it over, and we all agree that this was going to make and break us in ways we couldn't fathom. But we are humans, and this is what we do. The consequences are part of the action of adventure." With a smile, he breathed in and out then nodded to her and Thomas with confidence.

Kate smiled back proudly and replied, "Well then, now that we really don't have to worry about crew morale, how about the no alcohol policy, or limit it to only events?" The men chuckled and agreed that it would be a great idea, and they'd help implement it as soon as possible.

CHAPTER SIX

103 days from Earth, 07:00 hours

David Kim woke up slowly to his wall clock alarm, a soothing opera by Schubert—one of his random picks for morning shift wake up call. He was well rested from the eve of their departure festival with the Romeians, which was good, he thought to himself. He reminded himself that today he had the time off to go look at the new rifles to make scopes and laser red dots on them with some scrap duranium they brought on board. The Romeian security tactician had told them they could alter the outer chassis of the rifles and pistols to fit sights or optics, so it was an awesome plan, Morrison had said gleefully in his message. David just smirked and chuckled, thinking all security people were trigger-happy demolition junkies who had too much fun at times.

He got up and hopped in the five-minute shower pod and rinsed off. Then he jumped out and got dressed. He tucked his Epad into his shoulder pocket carrier and walked out the door to the mess hall for some breakfast chow before reporting to engineering and getting his orders from Elisa. David's quarters was across from the ship's infirmary, or sick bay, just near the aft section of the ship, adjacent to the cargo bay on the other side of the bulkhead. It was where the ship's hum was the loudest, yet it produced a soothing sound and feeling. That was the best part of this ship. When you went to sleep in your quarters, it was indeed a better sleep and relaxing feeling, unlike on Earth. This was now a known comfort among the crew, and the doctor even reported that this had improved the health of some crew members. Indeed, this was a good time for his fellow crewmates, once students at university on Earth. They'd pulled this plan off and were here to tell this story. The walk down the hall was quiet, with no one in sight. It was weird that at this time of morning other crew weren't also walking around, but when he walked around midship corner, he saw a massing of at least twenty crew in the observation deck, talking subtly about something. In a few seconds, he'd probably find out what they were discussing.

Thomas spotted him and was waving at him and Kim to join him erratically. As he walked up to the group he greeted them and then turned to Thomas, who excitedly showed him an Epad and explained, "Ok, so last night I was fooling around on my console, just listening for frequencies and zoning in and out of certain zones of space around us. Out there about one light year, I found an old Earth NASA satellite still pinging and floating around. Then I did a little more digging and found out it can be hacked to send out messages. Imagine piggy backing the signal with crew emails back home to family, telling them we're ok and what really happened!" Thomas was laughing excitedly, and David was now into this story "So I was thinking about asking the captain if we could do this, and if so, would you send a letter home?"

David was now feeling inspired, because he so wanted to tell his mom and dad and two sisters everything he had done and was a part of. He wanted to share all of this with them, and his heart yearned for them so much. The pit in his chest forced him to hold back a few tears as he replied, "You know what, let's make this happen. But first, I'm hungry, and I have to tend to a request from the security team. But yes, let's do it."

Kim wandered off toward the mess hall, where Cassidy was sitting at a table with two science officers, one being the chief astrometics officer, Marianne Perez, who was explaining something that had Cassidy laughing. He walked up to the table and sat down. The two ladies turned to him and asked, "Is it true a satellite was found and can be hacked to send messages home?"

David laughed and replied, "I'm told yes by the ship's comms officer himself." With a smile and shaking his head, he continued. "But before anything happens, I really am hungry and need a good breakfast!"

Cassidy got up and replied, "Right on! Here, my boy, let me bring you a plate of sausages and eggs with toast."

David was now happy and said, "Sure, thank you." The café owner went back into her kitchen and dished up some hot food then came back and served it. Kim dove in and loved it immediately. "Mmmmmm, scrambled eggs and buttered toast." She had dished a sausage and egg into folded bread to create a mini breakfast sandwich. He also had orange juice, which was quickly gulped down. Another plate put in front of him while, which this time he consumed a little slower.

"So are we able to hack an actual satellite, or is this a theory?" asked Perez.

David nodded and swallowed. "Well, I'm sure we have crew on board that can do what's being planned, if the captain thinks we can safely do it, which is another story." He bit his sandwich and swallowed before continuing. "Then there's the issue of whose back on Earth monitoring this satellite, and will they let these messages go through or will they try to discover what the signals are? Or maybe someone is watching them and could come out in an instant to see who's tampering with it, but who knows. All this is talk so far, and when the decisions are made, all will be told, I'm sure." He smiled and nodded then drank down some coffee and stared outside into space.

There was a quiet hum when a bunch of crew from the night shift walked in and lined up at the counter to dine before their rest period or off time. They looked happy and satisfied because they were now working on board a space vessel, something they'd only dreamed about. They were maintaining *Valkyrie* on shifts and working at consoles, looking at star maps, and sitting for long periods of time at the helm, but it wasn't mundane or tedious to them. This was their creation and they were keeping ship and crew alive to do what they'd dreamed of all their lives. He was a part of the crew and knew how his mates felt.

On his Epad was a routine matrix sensor analysis of the astrometrics lab. Then he stopped by security to look at the rifles and materials. He was going to help Thomas later with something about the satellite, which did pique his interest. Gulping down the last of his coffee, he said his farewell to Ms. Perez and indicated he might see her later or sooner, depending on the time it took to run the analysis. "Have a good day," he said. David got up and walked away as she looked up and replied, "See you later." He was anxious to get the day started.

Passing crew and smiling and greeting them, he continued through the observation deck, where Ahote was in one of his meditative states. When he passed the ship's doctor, Ricardo, he looked up and made his way over to David, calling out, "Mr. Kim, I have your heartburn medication ready for you whenever you can pick it up, or I can just leave it on the medicine counter in my office so you know where to get it; infirmary is open 24/7. If I'm not there I'm on rest period, so just ring me to my quarters from my desk console, or I can make house calls. Your choice! Ricardo was tall and handsome with olive skin. He was 6' tall with greyish white hair he dyed that way, as it made him seem a little older. He totally got the women, even though he was a gentleman who wanted the right woman to settle with and not just anyone. He was the kind that would go out of his way for people just because he was a thoughtful man. He was medium build, with green eyes and a moderate-sized nose and manly chiselled face. He was wanted by many women; he hated the fact that he had to turn them all down because he did feel guilty about one-night stands. He never again did that with anyone. He had stopped dating, and it was perfect for him to be a part of *Valkyrie* and distracted from romance until he could change his life situation. He wanted to see where *Valkyrie* took him and quit the town life for a bit.

David was happy to hear that his heartburn medication was ready, because sometimes too much coffee resulted in heartburn and drove him nuts. But all was good as the doctor walked with him to the lift and asked, "Is it true that there's a satellite somewhere out there crossing our path? I heard it was hackable." Kim laughed and reported that so far all he knew was the comm officer Thomas had found it and it was sending off pings, meaning it was still active and doing its thing. The doctor's face was excited and lit up at this news. While they walked, the doctor asked Kim about his day's schedule and then reported that he wanted to join the engineer for dinner because he had a few questions about a few female crew members David worked with. The doctor was interested in getting to know the small redheaded girl from Abbotsford and start a friendship perhaps, but he definitely needed some information about her before he invested his heart to another.

David's eyes went wide as he thought, *Wow, he's right when it comes to the matters of the heart, as you do have to protect yourself and your heart. It's easy to damage people and turn them away from a loving relationship.* "Well, I'll tell you what, Doc, give me a few days and I'll contact you with some info and we'll get you that date, if you can keep those heartburn meds coming." They nodded and shook hands and went their own ways, but Kim remembered the dinner invite and called out, "I get off duty at seven tonight. Come to the bridge lounge and kick back some non-alcoholic brewskies Thomas stocked and stored in the cargo hold before *Valkyrie*'s departure from Earth." Kim had to admit it was a great hiding place; he was tipped about it yesterday by another engineer who helped Thomas load the crates into the cargo bay. The doctor nodded and waved while he walked away, leaving David Kim to continue toward astrometrics. After a brief walk down the corridor, he entered the room with the main computer console and got to work scanning files and systems while monitoring the analysis showing up on his Epad, on which he saw all the cases and programs in the ship's computer systems. But today he was going to be going over the astrometrics systems and library files and navigational programs and running a scan on the information bytes library files to make sure all was intact and running at optimal levels. He hit a few buttons on the computer screen interface and began his half-hour scan of the system, letting it run while he walked over the 3D console, took a seat, and opened the alien races' files, starting with the Federation.

The Epad scanner went off thirty minutes later, indicating that all systems were fine and already backed up by another system that was downloaded onto a memory Epad, obviously the smart work of crewman Perez. She hadn't come back yet either. When he finished the task he shut the library program off and packed up to move on to the next task. "Shit!" he said in a hushed tone, remembering he had to go to engineering to check in with Torres for his shifts. He decided to call her on his Epad. David pulled the pad off the console and shut off the analysis program. He paged his chief engineer, who picked up with a cheery face.

"Well good morning to you, David. I noticed that your diagnostics analysis is complete. I've been monitoring your work progress since I noticed you didn't arrive to check in. I used the ship's internal scan system to track you down and decided to let you work. Obviously you got the memo."

He chuckled and continued. "Yes, sorry, I was distracted this morning after breakfast and proceeded straight to my first assignment. I'm now ready to go to the security bay and check in with the crew, take a look at those rifles, and run a few scans."

Torres smiled and nodded. "Right, well, don't let me hold you back. Good work on those scans, and let me know if you need me. I'll be in engineering all day."

They said their farewells, and David went back to packing up his gear then headed out of the office to go down a deck. He was thinking about assembling a team to help him overhaul the rifles, maybe three crew members would be good. He'd supervise and still be able to carry out his duties. David laughed out loud at his juggling act plan, thinking he just had to put names together and set up a lab for this project. Then it came to him: the engineers' lab! That room wasn't in use at the moment. Gleefully grinning, he turned the corridor and strode onto the lift. When the lift stopped, Mr. Kim continued down the corridor, passing a few more crewmen and nodding and greeting hello. He walked until he got to the back corridor, where security dug in their offices, and entered the main tactical room.

A few security crew were standing around their console watching the alien galaxy map and races file while discussing which alien race was more hostile. They were obviously studying the negative races who meant them harm. "Hey there, I'm the engineer requested to look at your rifles, to mount scopes."

Crewmen Kitchi turned to him with a smile and replied, "So you're the guy coming to look at our guns?" Sherri Wiles just laughed, leaving David a few seconds to clue in before he understood fully and blushed. The two girls giggled at his expense before the First Nations crewmen, Kitchi, spoke. "Follow me, engineer." They went into the next room, where he saw rifles on one wall and pistols on the other, neatly and snugly resting on individual racks. Kitchi was cute, he thought. She was at least his height, with dark brown braided hair. She was very pretty and a Canadian First Nations, which intrigued him deep inside. Her green eyes caught his attention because they were so shiny, and her smile was infectious with love and warmth all around. She seemed like that person you rarely run into, but when you do, they seem otherworldly.

They entered the weapons locker and she picked up a rifle and went over the specs they drew up of their scope idea and red laser sight. This made him chuckle and admit that he was impressed that security took their duty to secure and protect to heart when it came to a ship boarding, because most crew were disturbed by the idea. Kitchi gazed into his eyes and studied his expression while he went over the weapon. "It's a funny thought that on a ship of exploration," he said, "weapons are needed, but good people out for adventure have to arm themselves against people who are just rotten to the core." The women watched in silence as David explained how he could see an attachment rail to be assembled that would be an easy snap on and off. There was also the request for foregrips under the chassis, and maybe straps so one could carry it on their back. Of course, he'd have to make copies of the specs for the teams to go over, kind of the right fit deal.

The women were happy to hear that their concept was relevant and agreed that on and off attachments worked best for the security members. David looked up, smiled, and nodded farewell to the female security crew. He started out the door and down the corridor, thinking that it was a quick appointment that lasted only ten minutes, which now gave him a good thirty minutes to grab a coffee and plan his next assignment, which was helping Mr. Hilt hack into this satellite everyone was buzzing about. Suddenly the ship shook for a second, as if something had hit them at a high speed. The engineer was thrown off balance as he stumbled over his feet and nearly fell to the ground. The corridor was empty as code red lighting was activated, and David could hear shouts of surprise somewhere down one of the other crew ways. His Epad went off, and he looked at the screen to see Torres with a baffled look on her face.

"I need you in engineering asap!" He knew this had to be serious or of concern to be called into the main engine room like this. As he kept walking to the lifts, he saw two crew members, a man and a woman, leaning against a wall talking to their stations on an Epad, but all was well and the ship had not stopped yet. No abandon ship alarms were going off, so this was weird. The crew were ok, and David kept making his way to the lift that took him down to the bowels of the ship. As he turned the corner and walked straight ahead into the engineering room, he saw Torres and Tom Hilt standing at a console and looking at a vicinity chart of space outside and around them for a three-hundred-feet radius. They were discussing a possible cause of the shaking to be a leading edge of a super nova shockwave light years away, now at its weakest. Or it could have been space debris, but the ship's systems didn't show any objects around *Valkyrie* at the time of the jolt or before.

"What happened?" asked David as he walked up to the two crewmen. He watched the screen and saw a very small anomaly signature resonance signal on the display quickly pass by. "There, right there," he called out. "That's a resonance signal coming from a tiny anomaly we just passed."

Mr. Hilt looked up and spoke. "Right, but it's what came out of the anomaly I need fixing up, my friend, and I do believe you are that man." Thomas pushed a few buttons, and a message wave popped up on screen with fluctuating frequencies, lots of static, and what sounded like someone calling out in a deep tone, maybe a woman's raspy voice. The three of them knew now that they had stumbled upon a subspace message, but the next question was what was the message, who sent it, and where from. The crew were silent as Torres replayed the sensor logs for the ship five minutes before the shake and after, just to make sure. Sure enough, the same conclusion was reached. *Valkyrie* had run across an anomaly too small for the ship's systems to pick up on, but it had a kick to it when the ship ran through it. Now was the matter of the message. Torres played it again to only hear garble and static. David hit a few keys to clean up the static and buffer the message strength to the normal decibels humans were used to. Then a few algorithm scans and aligning the signal with the ship's computers to run the message through a filter. Two minutes later, a woman's voice was clearly heard. The surrounding crews stopped and looked toward David and the console to hear a voice call out:

"This is Captain Jordan Winston of the Terran Federation Alliance calling to the vessel *Valkyrie*. The tear between our space is perfect to hide this message for you. This is a warning call. The Draconians are scattered and have surrounded this area of space, cloaked by sitting in subspace pockets looking for human vessels. Be very aware. Right now we're dispatching Pleiadean fleets to come escort you. We are not your enemies; do not fear us. But be careful, guys. You are most likely being watched. Jordan out!"

"Ok, what in the heck?" Torres spoke out loud, looking up to Thomas. She backed away from the console, whispering in a hushed and shocked tone. "What did she just say?" She continued stammering her words. "You're the comms guy; you got this message on board. How are we going to proceed with this?"

Just then the captain walked in. David turned around, now baffled and speechless from what he had heard, while his heart pounded out of amazement. "A message from the Terran Federation directed to us, Captain." His face showed shock as the captain walked up to the console, demanding a report and overhearing David Kim's conversation. Thomas was able to chime in and reported that *Valkyrie* was flying along, smooth sailing, when out of nowhere came a shudder. The ship had hit something, but the helm next to him saw nothing in their path. Then David Kim was called and figured out it was an anomaly they'd hit. Now it appeared to be a tear in space with a small hole to another universe.

Kate Braxton stood there, her expression bewildered and eyes wide like dinner plates. She was even more frozen in disbelief when she heard the message herself, her voice and all. Kate turned to the crowd, mouth open, while she shook and pointed her finger toward the console. "To think Draconian ships are surrounding us and hiding in subspace pockets, and the Terrans are hiding a message in a tear in space in the multiverse on the other side."

She looked back and put her hands on her hips. Then she asked for a damage report on the ship. Another engineer reported that *Valkyrie* was ok and there was no damage to the hull or engines. All systems were running smoothly. Kate looked back at the console and ordered the coordinates to be logged and sent to astronomics to be analyzed and charted for the ship's records. With that, she called Kim and Hilt to follow her to the bridge deck briefing room for a report.

Meanwhile, the engineering crew were ordered to keep the engines going, as they were on a mission and a deadline to meet up at Pleiadean. There they could report the message and event to the Pleiadean space committee, or whoever oversaw space travel affairs. The captain walked back toward the bridge in a hurried pace, just too excited to hear the report she was going to be given, but she remembered that she was supposed to do something. Catching herself, she turned around and motioned the comms officer to a chair and took a seat herself

In the briefing room, Kate sat listening to the stories shared by the bridge crew. She was shown external scans taken during the shudder. Sure enough, Kate pointed out the energy signature that showed an electro anomaly occurring in that sector, but it was too small to see or pick up at that moment in time. "I agree with your explanation, Kim. The question is, where are these hidden Draconian ships that Jordan Winston is talking about, and where is he right now? Mr. Hilt, I want you to do what you do best when you're bored at the comms station."

David explained that the multiverse theory was somewhat proven, but there were skeptics who said it can't exist, and they had good points. "It's a toss-up in the science community, but Ahote, are there not stories from your peoples and others about interdimensional portals of energy or stargates opening up in certain ancient, holy places where people disappear in plain sight?"

Ahote smiled and chuckled. "Yes, David, there are, but those are stories, and scientists have found energy portals there."

Ecstatically, David replied, "Yes, and they said all these stories of Pleiadeans were stories too, and now we know they're real and wormholes! We know they exist. Where do those portholes at those sacred sites go? Other dimensions? Parallels? The Multiverse Theory!"

"So my other question is: Who are these Terran Federation Alliance we were told about?" Kate openly pondered.

Thomas added, "I did some research on them before, Captain, and they're in our computer files the Romeians sent us. They're a small faction of Earth human rebels or freedom fighters who got angry and left those Earth secret space thugs we ran into. They were tired of abusing their own people while letting the Draconians infiltrate our institutions, so they broke away and went AWOL, started an Earth Man thus Terran, a federation of them and other humans from other worlds who sympathize with the Terran plight. With an aggressive nature, they attack any and all alien races, non-human, who stand by and watch or are part of human abuses in our galaxy. They are violent and very xenophobic."

The four paused and looked at each other before dispersing. The captain compiled the report into her computer console and sent the other two back to their stations. Kate had to admit that this was a lot of information to think on, and she needed a coffee and someone to throw ideas at. *Cassidy*, she thought with a smile of comfort. On the way down, Kate ran into another female crewmember of the sciences department who assisted in astronomics. She was on her way to the mess hall for lunch. The captain smiled as she passed by and was cut off by Kyle on the bridge on her Epad.

"Captain and all bridge officers to the bridge please. You're going to want to see this." Alert from the events that just unfolded, Kate immediately replied that she was on her way. They strolled out into the corridor together, thanking the good "whoever" for putting in the night hours mood lighting at 3:00 a.m. The crew laughed while slowly pussy footing it out and down the corridor and onto the bridge, where proximity alarms were going off on the communications and pilot's console.

"What the hell?" Thomas replied and then was cut off by Mr. White, who reported that they were getting massive ion readings just half a mile up ahead and closing in. He ordered the ship to slow down as they were doing now; he explained that five different ion fluctuations were seen by ship's sensors, and they all were fairly similar to the same frequency as the Earth secret spaceship. It was weird because the signatures were booming in and out and appearing one by one. Now they were sitting stationary.

"How long till we can see them, Ms. Blackwood?" Kate asked with a curious furrow on her face as she walked back toward the pilot's console. She was interrupted by Thomas's reaction as he threw off his headset, screaming out and holding his ears in pain as he fell to the floor. Kate looked at him and ordered the ship's doctor to be summoned immediately. With concern, she turned to Kyle, and the whole bridge broke out in a loud, high pitch, scratchy squelching sound that made everyone cover their ears in discomfort and try to make their ways away from the bridge. Ahote helped drag Thomas into the corridor while the high pitch sound died down after what seemed like a few minutes. Then they heard what sounded like a voice in some distorted radio transmission from an old radio or something: "No harm ... stop ... please."

Kate ran back to the helm and noticed that a bunch more blips were showing up in a wall pattern blocking the *Valkyrie*'s flight path and causing her to come to the conclusion that if she kept going, she would collide with whoever was out there. Secondly, she wouldn't be able to go around them, as they seemed to have erratic manoeuvres on her pilot's screen. Kate saw the screen was picking up a massive fleet spanning larger than Manhattan north and south and heading slowly toward them. Ahote walked back while watching the windshield with a light sweat about him and a worried look on his face. Then that all changed when a bunch of white and blueish lights could be made out in the distance. They appeared to be getting large and closer. Kate counted ten lights side by side with more still appearing directly off their bow a few hundred metres away. "Do you believe them? I mean the message?" Kate spoke and turned to face the commander.

"Well, we have made it this far, made first contact, and are just weeks away from the Pledan system. We've made our presence known." She turned back to her console and punched up a manual greeting selection and typed in the text: "We are stopped but ready to defend ourselves if need be! Who are you?" The captain and commander hovered over the console, waiting for a reply but knowing it was a blind shot in the dark. But it was something to try for the time being. Kate sent it out to the lights that now were a hundred times larger and in front of them, blocking off all avenues of escape. "Let's hope this isn't a bigger guy we need to worry about. Get security up here immediately." She walked back to the seat console and paged the security team to come to the bridge right away.

Moments later, the bridge crew were back on the bridge at their posts going over the data and frequencies on their monitors and trying to save and store the information in the ship's data banks for future reference. Kate and the commander talked over the greeting procedure again and what they should do if it's an ambush. She had stated that they could have security on standby and ready to go with their new weapons if necessary, or they could just wait and see what happened; after all, they hadn't attacked or shown aggressive tendencies toward *Valkyrie* thus far, so maybe they really were benevolent.

Ahote nodded in agreement then sat back in his chair with his arms crossed. He tapped his chair Epad and looked up their position in relation to the Pledan system. They were pretty close, as he could see the purple cloud much brighter now, and a few stars appeared. "We're close, by golly, Captain. Maybe you're right. We should wait this one out, but be ready, no doubt." Kate looked at him and smiled before a light flashed. The ships were now a few hundred feet off of *Valkyrie*'s bow, sitting stationary and quiet.

The bridge crew sat silently, staring at the sight of these huge, smooth, sliver, boxy looking cruisers glaring down on them like mountains to a climber, yet no malicious or hostile acts were made at this point. The ships were covered with a soft, bright whitish blue glow, and it was hard to make out their shape. But if Kate had to guess, they'd probably be something like a boxy-edged turd shape.

"Holy shit! You don't see this every day!" Mr. Gagnon was sitting in the engineer's console chair and staring up in disbelief, his mouth open and jaw dropped as he mumbled. "I … I don't not believe in them, but still have a skeptic's mind. Holy crap, they exist." His heart was pounding harder and harder, not knowing whether to feel nervous or excited. The atmosphere on the bridge was quiet and not quite tense, but still vigilant yet somewhat at ease for a reason Kate couldn't explain. A bright ball of blue light appeared suddenly on the bridge and hovered above the floor, getting everyone's undivided attention. Then a male disembodied voice was heard by the bridge crew: "Welcome, crew of the *Valkyrie*. You have no reason to fear us, nor do we mean you harm, but we suggest you come with us as we escort you back to the Pleiades star base. Congratulations, children, you have made it safely, and we will speak very soon. Prepare to be beamed aboard our vessel when you're ready. Welcome to the Galactic Federation on Pleiadean Prime."